Audio IC projects

The Maplin series

This book is part of an exciting series developed by Butterworth-Heinemann and Maplin Electronics Plc. Books in the series are practical guides which offer electronic constructors and students clear introductions to key topics. Each book is written and compiled by a leading electronics author.

Other books published in the Maplin series include:

Computer interfacing	Graham Dixey	0 7506 2123 0
Logic design	Mike Wharton	0 7506 2122 2
Music projects	R A Penfold	0 7506 2119 2
Starting Electronics	Keith Brindley	0 7506 2053 6

Audio IC
Projects

Edited by Keith Brindley

Newnes

An imprint of Butterworth-Heinemann Ltd
Linacre House, Jordan Hill, Oxford OX2 8DP

$\mathcal{R}$ A member of the Reed Elsevier group

OXFORD   LONDON   BOSTON
MUNICH   NEW DELHI   SINGAPORE   SYDNEY
TOKYO   TORONTO   WELLINGTON

British Library Cataloguing in Publication Data
A catalogue record for this book is available from the
British Library
ISBN 0 7506 2121 4

Library of Congress Cataloguing in Publication Data
A catalogue record for this book is available from the
Library of Congress

Edited by Co-publications, Loughborough
Typeset and produced by Sylvester North, Sunderland
all part of The Sylvester Press
Printed in Great Britain by Clays Ltd, St Ives plc

Contents

Preface

This book is a collection of projects previously published in *Electronics — The Maplin Magazine*. In their original guise they formed part of the Data File series so popular with regular readers.

Each project is based around an integrated circuit device, selected for publication because of its special features, because it is unusual, because it electronically clever, or simply because we think readers will be interested in it. Some of the devices used are fairly specific in function — in other words, the integrated circuit is designed and built for one purpose alone. Power amplifier integrated circuits are good examples. Others, on the other hand, are not specific at all, and can be used in any number of applications. Bucket brigade delay lines, say, can be used in all sorts of audio projects. Naturally, the circuit or circuits associated with each integrated circuit device reflect this.

While all circuits given with all integrated circuits here are intended for experimental use only — they are not full projects by any means — a printed circuit board track and layout are detailed. To help readers, the printed circuit boards (and kits of parts for some of the projects, too) are available from Maplin, but — as these circuits *are* for experimental use only — all constructional details and any consequent fault finding is left upto readers.

This is just one of the Maplin series of books published by Newnes books covering all aspects of computing and electronics. Others in the series are available from all good bookshops.

Maplin Electronics Plc supplies a wide range of electronics components and other products to private individuals and trade customers. Telephone: (01702) 552911 or write to Maplin Electronics, PO Box 3, Rayleigh, Essex SS6 8LR, for further details of product catalogue and locations of regional stores.

Projects by integrated circuit device number

1 Power amplifiers

Featuring:

LM386 power amplifier

The LM386 is a general-purpose power amplifier designed for use in low voltage applications. Voltage gain is internally set to 20 (26 dB) to keep the number of external parts to a minimum. However, the addition of an external resistor and capacitor between pins 1 and 8 allows voltage gain to be increased to any value up to 200 (46 dB). When operating from a 6 V supply the quiescent power drain is only 36 mW making the LM386 particularly useful for battery operation.

Application hints

To make the LM386 more versatile, two pins (1 and 8) are provided for gain control. If pin 1 and pin 8 are left open, as in Figure 1.1, the internal 1.35 kΩ resistor sets the voltage gain to 20 (26 dB). By connecting a capacitor from pin 1 to pin 8, as in Figure 1.2, bypassing the internal resistor, the voltage gain can be increased to 200 (46 dB). Voltage gain may be set to any value between the limits of 20 and 200 by connecting a resistor in series with the capacitor, as shown in Figure 1.3. Gain control may also be achieved by capacitively coupling a FET or resistor from pin 1 to ground.

It is possible to connect external components in parallel with the internal feedback resistors to tailor the gain and frequency response for different applications; for example the bass response may be effectively increased by connecting a capacitor and resistor in series between

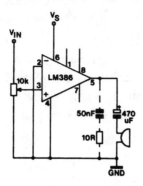

Figure 1.1   Amplifier with gain of 20 (26 dB)

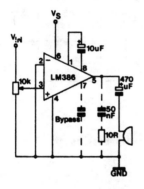

Figure 1.2   Amplifier with gain of 200 (46 dB)

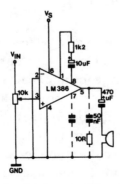

Figure 1.3   Amplifier with gain of 50 (34 dB)

pin 1 and pin 5, as in Figure 1.4. A resistor value of 15 kΩ will produce an effective bass boost of approximately 6 dB. The lowest resistor value for stable operation is 10 kΩ if pin 8 is open because the amplifier is only compensated for closed loop gains greater than 9; values as low as 2 kΩ can be used if pin 1 and pin 8 have been bypassed.

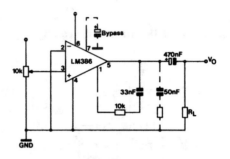

Figure 1.4 Amplifier with bass boost

Power supply requirements

The LM386 will operate over a wide range of voltages between 4 V and 12 V making it ideal for battery operation, the optimum voltage for minimum distortion being around 6 V. If the LM386 is used with a mains derived d.c. power supply it is important that the supply rail is adequately decoupled to prevent the introduction of mains-derived noise into the amplifier. Decoupling close to the integrated circuit is also necessary, to prevent any high frequency instability.

Input biasing

Referring to the integrated circuit schematic, Figure 1.5, it may be seen that both inputs are biased to ground with a 50 kΩ resistor. The base current of the input transistors is around 250 nA, so the inputs are at approximately 12.5 mV when left open. If the d.c. source resistance driving the LM386 is higher than 250 kΩ there will be very little additional d.c. offset. Where the d.c. source resistance is less than 10 kΩ the unused input can be shorted to ground to keep the offset low. For d.c. source resistances between these values any excess offset may be eliminated by connecting a resistor equal in value to the d.c. source resistance, between the unused input and ground. When using the LM386 with higher gains it is necessary to bypass the unused input to prevent degradation of gain or any possible instability; this may be achieved by connecting a 0.1 µF capacitor or a short (depending on the d.c. source resistance on the driven input) from the unused input to ground.

Printed circuit board

A high quality fibreglass printed circuit board, with printed legend is available as an aid to construction of the basic LM386 amplifier circuit. Referring to Figure 1.6, the power supply is connected between P1 (+V) and P2 (0 V), the optimum voltage being around 6 V (see Table 1.1). Input signals are applied between P3 and P4 and the output is taken from P5 and P6 (the amplifier will operate satisfactorily into an 8 Ω load). The overall volt-

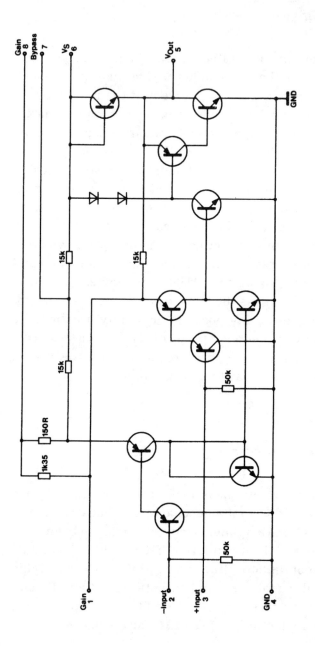

Figure 1.5 IC schematic diagram

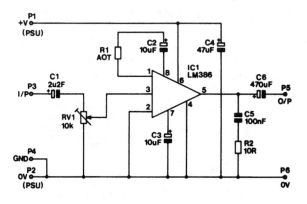

Figure 1.6   Module circuit diagram

Parameter	Conditions	Min	Typ	Max
Supply voltage (V_s) d.c.		4 V d.c.		12 V
Quiescent current (I_q)	V_s = 6 V, V_{in} = 0 V		4 mA	8 mA
Input resistance (R_{in})			50 kΩ	
Output power (P_{out})	V_s = 6 V, R1 = 8 Ω, THD = 10%	250m W	325 mW	
Voltage gain (A_v)	V_s = 6 V, f = 1 kHz		26 dB	
	10 μF capacitor from pin 1 to pin 8 of IC			46 dB
Bandwidth (BW)	V_s = 6 V, pin 1 and pin 8 of IC open		300 kHz	
Total harmonic distortion (THD)	V_s = 6 V, R1 = 8 Ω, P_{out} = 125 mW, f = 1 kHz pin 1 and pin 8 of IC open		0.2%	
Input bias current (I_{bias})	V_s = 6 V, pin 2 and pin 3 open		250 nA	

Table 1.1   Electrical characteristics of LM386

Audio IC projects

age gain of the amplifier is set by preset RV1, the maximum gain determined by the value of resistor R1. Figures 1.7 to 1.9 show various applications and Figure 1.10 gives details of various functions of the LM386.

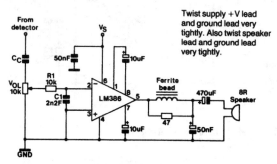

Figure 1.7 Power amplifier for AM radio

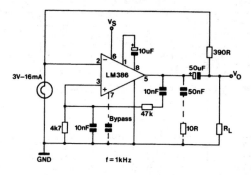

Figure 1.8 Low distortion power wienbridge oscillator

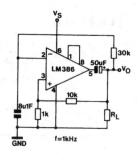

Figure 1.9 Square wave oscillator

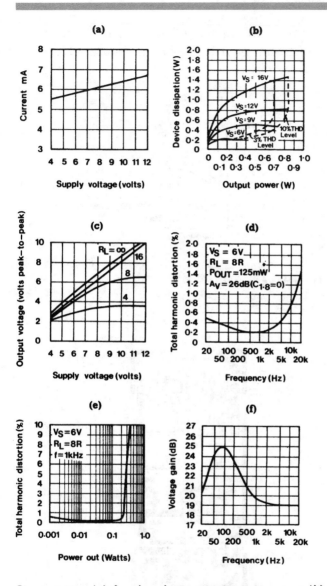

Figure 1.10 (a) Supply voltage vs quiescent current (b) device dissipation vs output (8 Ω load) (c) peak to peak output voltage swing vs supply voltage (d) distortion vs frequency (e) distortion vs output power (f) frequency response with bass boost

Audio IC projects

Pinout information is given in Figure 1.11. The layout of the printed circuit board is shown in Figure 1.12.

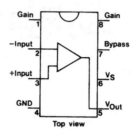

Figure 1.11   LM396 pinout

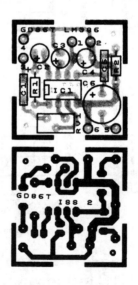

Figure 1.12   PCB overlay and track

LM386 board parts list

Resistors — All 0.6 W 1% metal film

R1	select on test	1	
R2	10 Ω	1	(M10R)
RV1	10 k vert enclosed preset	1	(UH16S)

Capacitors

C1	2μ2F 35 V tantalum	1	(WW62S)
C2,3	10 μF 50 V PC electrolytic	2	(FF04E)
C4	47 μF 16 V min elect	1	(YY37S)
C5	100 nF 16 V minidisc	1	(YR75S)
C6	470 μF 16 V PC electrolytic	1	(FF15R)

Semiconductors

IC1	LM386N-1	1	(UJ37S)

Miscellaneous

constructor's guide	1	(XH79L)
printed circuit board	1	(GD86T)
1 mm PCB pin	1 pkt	(FL24B)

LM1875 20 watt audio amplifier

The LM1875 is a general purpose audio power amplifier that offers very high quality output using a minimum of external components. The integrated circuit pinout is shown in Figure 1.13. The device operates over a wide range of power supply voltages from 20 V to 60 V d.c. and will deliver 20 watts r.m.s. into a 4 Ω or 8 Ω load when operated from a 50 V supply. If a 60 V supply is used, output powers up to 30 W r.m.s. may be produced (if an increase in distortion is acceptable). By using advanced circuit techniques the amplifier integrated circuit offers minimal distortion even at high power levels. Other features include wide bandwidth, high gain, large output voltage swing and overload protection. Table 1.2 gives the electrical characteristics of the LM1875 and Figure 1.14 shows some typical performance characteristics of the device.

Stability and distortion

The LM1875 is designed to be stable when operated with a closed loop gain greater than ten; however, as with any other high current amplifier, it may oscillate under certain conditions. Oscillation is often caused by poor circuit board layout or associated with input/output connections. When designing a layout it is important to return the load earth and the signal earth to the main earth point via separate paths. Preferably the load earth should be connected directly to the 0 V terminal of the

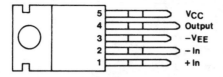

Figure 1.13   Pinout of the LM1875

power supply. If the input and load earths are connected to 0 V via the same rail, high currents on the rail can generate voltages which effectively act as input signals, leading to high frequency oscillation or distortion. It is recommended that the earth (0 V) rails are kept as short as possible and that decoupling capacitors and output compensation components are kept close to the integrated circuit to minimise the effects of track resistance and inductance. Sometimes oscillation can be caused by

Parameter	Conditions	Typical	Tested limits
Supply voltage			60 V (±30 V) max
Supply current	$P_{out} = 0$ W	70 mA	100 mA
Output power (P_{out})	THD = 1%	25 W	
Load impedance			4 Ω–8 Ω
THD	$P_{out} = 20$ W, 4 Ω load,		
	$F_o = 1$ kHz	0.022%	
	$P_{out} = 20$ W, 4 Ω load,		
	$F_o = 20$ kHz	0.07%	
Full power bandwidth			d.c.–250 kHz (–3 dB)
Open loop gain	d.c.	90 dB	
Max slew rate		8 V/μs	

Table 1.2   Electrical characteristics of the LM1875

Audio IC projects

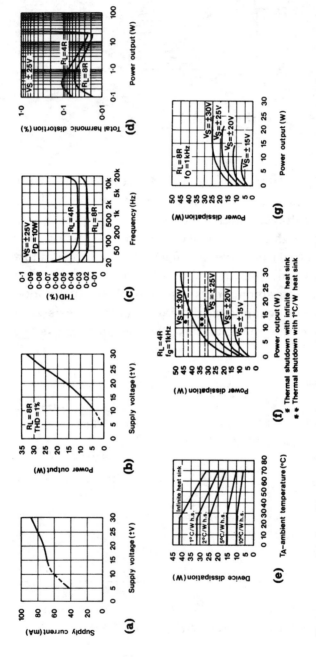

Figure 1.14 (a) Supply current vs supply voltage (quiescent) (b) power output vs supply voltage (c) THD vs frequency (d) THD vs power output (e) device dissipation vs ambient temperature (f) power dissipation vs power output (4 ohm load) (g) power dissipation vs power output (8 ohm load)

stray coupling between the output and input leads, especially if the leads are long and the source impedance is high; in order to avoid this, these leads should be kept as far apart and as short as possible. It is often possible to prevent oscillation due to stray input/output coupling by fitting a 50 pF–500 pF capacitor across the circuit input terminals.

In addition to preventing problems with spurious oscillation, layout can also be an important factor in achieving minimum distortion. For low distortion the power supply wiring is also important; this should be kept as far away as possible from the input wiring to help prevent non-linear power supply currents being induced into the integrated circuit inputs. If possible the power supply wires should be kept perpendicular to the circuit board for a few centimetres.

Thermal protection and heatsinking

The LM1875 incorporates a sophisticated thermal protection system to help prevent any long term thermal stress to the device. If the integrated circuit's die temperature reaches 170°C the amplifier shuts down until the temperature drops back to around 145°C; if however temperature starts to rise again the device will then shut down at around 150°C. The effect of the above characteristic is to allow the device to rise to a relatively high temperature under short duration fault conditions but limit the temperature of the device if the fault condition is sustained; this helps to improve the long term reliability of the integrated circuit.

Audio IC projects

It is important that the amplifier is always operated with a heatsink because even when off load, the device may dissipate up to 6 W and when on load the dissipation may be as high as 30 W. A heatsink should be chosen that is sufficient to keep the temperature of the device well below shutdown temperature. For reliability the heatsink should be the largest possible for the space available.

If the amplifier is powered from a single rail supply, the integrated circuit mounting tab may be bolted directly to the chassis (0 V). When the device is powered from a split rail supply, to avoid damage, it is important that the tab is completely isolated from 0 V; an insulating bush and a mica washer is usually used for this purpose. If the amplifier is powered from a split supply, a larger heatsink may be necessary because the thermal connection to the heatsink through a mica washer is less efficient than a direct connection

Current limit

In addition to thermal protection, the LM1875 also provides current limiting. A power amplifier can be easily damaged by excessive applied voltage or current flow. Reactive loads are often a problem due to the fact that they can draw large currents at the same time as high voltages appear on the amplifier's output transistors. To prevent any damage that may occur, the LM1875 limits the current to around 4 A and also lowers the value of this current limit when high voltage appears across the output of the device. Protection is also provided against the excessively high voltages that may appear on the output of the device when driving non-linear inductive loads.

Power supply

The amplifier may be powered from either a single or split rail supply and will operate over a wide range of voltages between 20 V and 60 V (between ±10 V and ±30 V when powered from a split rail supply). Current requirements depend very much on output power and may range from a few mA to over 1 A. It is important that the power supply is adequately decoupled to prevent the introduction of mains derived noise into the amplifier.

Printed circuit board

A high quality fibreglass printed circuit board with printed legend is available for the basic LM1875 audio amplifier application. Two different versions of the amplifier may be constructed using the same printed circuit board; one version is for use with a single rail supply (see Figure 1.15) and the other is for use with a split rail supply (see Figure 1.16). A combined circuit diagram of both versions of the amplifier is shown in Figure 1.17 for reference purposes; this is the circuit used to produce the printed circuit board which is shown in Figure 1.18. Provision is made for a printed circuit board mounted heatsink bracket; power should not be applied to the

Audio IC projects

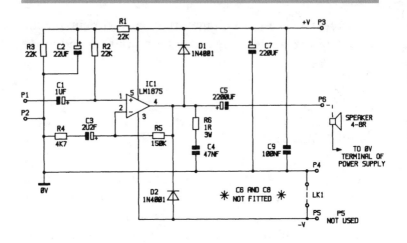

Figure 1.15 Amplifier for single rail supply

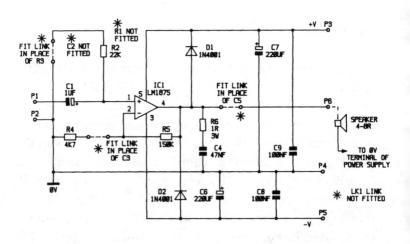

Figure 1.16 Amplifier for split rail supply

18

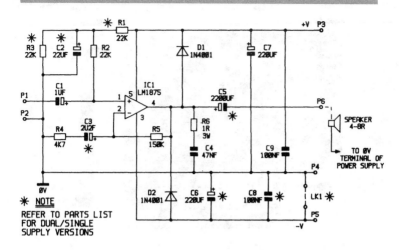

Figure 1.17 Combined circuit to which PCB is designed

amplifier until the bracket has been bolted securely to a suitable heatsink (for example, a heatsink with at least 1500 cm^2 surface area). Please note that if the amplifier is powered from a split rail supply, the integrated circuit tag must *not* be electrically connected to the chassis (0 V); an insulating bush and a mica washer must be used to isolate the tag from the heatsink (see Figure 1.19). It is recommended that heat transfer compound is smeared between the device and the heatsink to facilitate the conduction of heat away from the device. If a mica washer is used, the compound should be applied on both the integrated circuit and heatsink sides of the washer. A larger heatsink may be necessary for the split rail version of the amplifier.

19

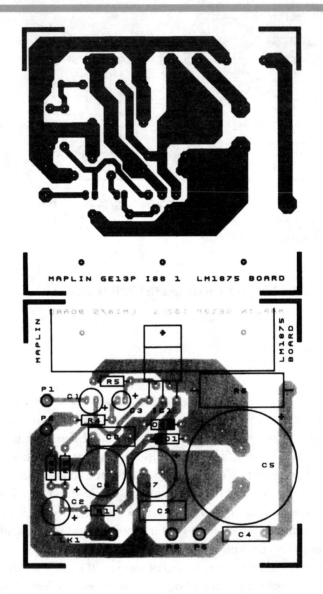

Figure 1.18   Track and layout of PCB

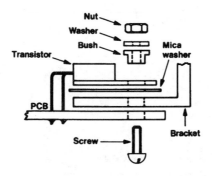

Figure 1.19   Mounting IC1

For connection information, refer to Figure 1.15 or Figure 1.16 as appropriate. The power supply is connected to P3(+V), P4(0 V) and P5(–V); if the amplifier is powered from a single rail supply P5 is not used. Heavy gauge wire should be used for the power supply and output connections and all leads should be kept as short as possible. The signal input is applied between P1 and P2 using screened cable (XR12N) and the output is taken from P6, the load earth being connected directly to the 0 V terminal of the power supply. Finally, Table 1.3 gives the specification of the prototype amplifier that was built on the printed circuit board.

Power supply voltage	20 V–60 V d.c.	(±10 V–±30 V d.c.)
Power supply current	(quiescent)	85 mA (at 60 V)
Voltage gain	(set by value of R4 and R5)	30 dB
Full power bandwidth	(4 Ω load)	20 Hz–250 kHz (–3 dB)
Output load impedance		4 Ω to 8 Ω

Table 1.3   Specification of prototype

Single rail parts list

Resistors — All 0.6 W metal film

R1,2,3	22 k	3	(M22K)
R4	4k7	1	(M4K7)
R5	150 k	1	(M150K)
R6	1 Ω 3 W wirewound	1	(W1R)
LK1	link fitted		

Capacitors

C1	1 µF 100 V PC electrolytic	1	(FF01B)
C2	22 µF 63 V PC electrolytic	1	(FF07H)
C3	2µ2F 100 V PC electrolytic	1	(FF02C)
C4	47 nF polyester	1	(BX74R)
C5	2200 µF 63 V PC electrolyic	1	(JL29G)
C6	not fitted		
C7	220 µF 63 V PC electrolytic	1	(FF14Q)
C8	not fitted		
C9	100 nF polyester	1	(BX76H)

Semiconductors

IC1	LM1875N	1	(UH78K)
D1,2	1N4002	2	(QL74R)

Miscellaneous

printed circuit board	1	(GE13P)
bracket	1	(YQ36P)
1.3 mm PCB pins	1 Pkt	(FL21X)
M3 bolt × 12 mm	1 Pkt	(BF52G)
M3 nut	1 Pkt	(JD61R)
M3 washer	1 Pkt	(JD76H)

Split rail parts list

Resistors — All 0.6 W metal film

R1	not fitted		
R2	22 k	1	(M22K)
R3	linked out		
R4	4k7	1	(M4K7)
R5	150 k	1	(M150K)
R6	1 Ω 3 W wirewound	1	(W1R)
LK1	link not fitted		

Capacitors

C1	1 µF 100 V PC electrolytic	1	(FF01B)
C2	not fitted		
C3	linked out		
C4	47 nF polyester	1	(BX74R)
C5	linked out		
C6,7	220 µF 63 V PC electrolytic	2	(FF14Q)
C8,9	100 nF polyester	2	(BX76H)

Semiconductors

IC1	LM1875N	1	(UH78K)
D1,2	1N4002	2	(QL74R)

Miscellaneous

printed circuit board	1	(GE13P)
bracket	1	(YQ36P)
1.3 mm PCB pins	1 Pkt	(FL21X)
M3 bolt × 12 mm	1 Pkt	(BF52G)
M3 nut	1 Pkt	(JD61R)
M3 washer	1 Pkt	(JD76H)
mounting kit	1	(WR23A)

TDA7052 1 watt power amplifier

The TDA7052 is a 1 watt mono amplifier which is ideal for use in low power battery operated or similar equipment. Because it requires very little in the way of external components to function, the device is ideal for use in portable apparatus. Figure 1.20 shows the integrated circuit pinout and Table 1.4 lists some typical electrical characteristics for the device. A block diagram of the integrated circuit is shown in Figure 1.21.

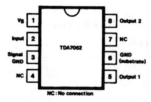

Figure 1.20 IC pinout

Circuit description

The device makes use of the bridge tied load principle allowing relatively high power to be developed into an 8 Ω load at low voltages. Using this method it is possible to achieve output powers up to 1.2 W into 8 Ω at 6 V with 10% distortion. In addition to power supply voltage, temperature also plays an important part in setting the maximum power limit for the device. The package is capable of dissipating higher power at lower temperatures and this is illustrated by the power derating curve shown

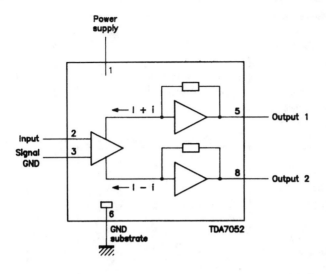

Figure 1.21   IC block diagram

in Figure 1.22. Amplifier gain is internally set to approximately 40 dB making the device suitable for direct amplification of comparatively small signals to a suitable level to drive a loudspeaker directly. Input impedance is typically around 100 kΩ; input attenuators are fairly easy to implement.

Power supply requirements

The TDA7052 will operate over a wide range of power supply voltages between 3 V and 15 V. As with all amplifiers, it is important that the power supply is properly de-coupled at both high and low frequencies; this prevents any mains derived noise being introduced into the

Audio IC projects

Parameter	Conditions	Min	Typ	Max
Supply voltage		3 V	6 V	15 V
Supply current	Quiescent, load disconnected	4 mA		8 mA
Output power	THD = 10%, 8 Ω load		1.2 W	
Voltage gain		39 dB	40 dB	41 dB
Total harmonic distortion (THD)	Output power = 0.1 W		0.2%	1.0%
Frequency response		20 Hz		20 kHz
Output offset voltage	Source impedance (R_s) = 5 kΩ		100 mV	
Input impedance			100 kΩ	
Input bias current			100 nA	300 nA
Storage temperature		−65°C		+150°C

Table 1.4 Electrical characteristics of TDA7052 — the above specification applies to: power supply = 6V , output load = 8 Ω, signal frequency = 1 kHz, ambient temperature = 25°C unless otherwise noted.

system and also helps to reduce the possibility of instability. Current consumption is dependent on the load being driven and the amount of power developed. The quiescent current is typically around 4 mA. Low voltage operation and the requirement of very few external components makes the TDA7052 ideal for battery-powered equipment. The type of battery used will be determined by the maximum current consumption under full load and also of course by the size of the equipment into which the amplifier is to be incorporated.

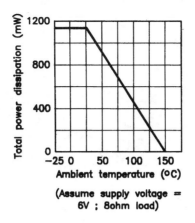

(Assume supply voltage = 6V ; 8ohm load)

Figure 1.22 Power derating curve

TDA7052 kit

A kit of parts including a high quality fibreglass printed circuit board is available as an aid to constructors, for a simple application circuit using the TDA7052. Figure 1.23 shows the circuit diagram of the module and Figure 1.24 shows the printed circuit board layout.

For wiring information refer to Figure 1.25. Printed circuit board pins, P5, P6 and P7 provide for the connection of a rotary potentiometer volume control (RV1) and this component may either be soldered directly to the pins or connected by a short run of cable as appropriate. Long lengths of cable should not be used for this purpose because the amplifier has a high input impedance and the use of long leads at the input could cause instability problems.

Audio IC projects

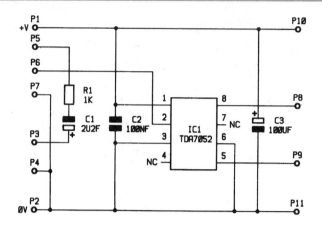

Figure 1.23   Circuit diagram

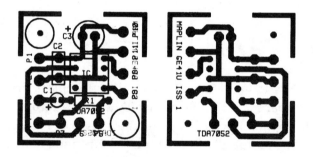

Figure 1.24   PCB legend and track

A 3 V to 15 V power supply that is capable of delivering at least 400 mA is required to power the circuit. If a mains derived d.c. power supply is used it is important that this is adequately de-coupled to prevent the introduction of low frequency noise (mains hum) into the system. Power supply connections are made to P1(+V) and

P2(0 V). It is a good idea to use screened lead for input connections to the amplifier module as this helps prevent pickup of external noise and reduces any stray coupling between the input and output; this is particularly important where low level signals are involved. Input signals are applied to P3(i/p) and P4(0 V) and output (loudspeaker) connections are made to P8 and P9. Printed circuit board pins P10(+V) and P11(0 V) are provided for additional power supply connections to auxiliary

The overall gain of the module is set by the values of resistor R1 and potentiometer RV1 which act as a potential divider network, attenuating the input signal. With the component values supplied in the kit, the gain of the circuit with potentiometer RV1 set at maximum is approximately 39 dB. Optimum performance and maximum power is achieved when the amplifier is operating into an 8 Ω load; however, higher impedance loads may be used with a reduction in output power. A suitable loudspeaker for general purpose use is Maplin stock code YT25C.

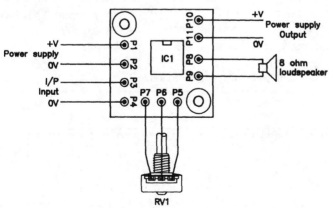

Figure 1.25   Wiring diagram

Applications

Being a general purpose module, the TDA7052 1 watt power amplifier is suitable for use in many different applications where simple but effective audio power amplification is needed. Typical uses for the module could include low power audio amplification in portable radios, cassette recorders and related devices. Also, the high gain capability of the circuit makes it ideal for use in intercoms and baby alarms, where the module may be used to amplify signals from a microphone with very little pre-amplification to a suitable level to drive a loudspeaker directly.

Finally, Table 1.5 shows the specification of the prototype TDA7052 1 watt power amplifier module.

Power supply voltage range	3 V–15 V
Power supply current (quiescent)	4 mA at 6 V
Power supply current (maximum)	340 mA at 6 V (For output power = 1 W into 8 Ω load)
Total harmonic distortion (at 1 kHz)	0.7% at 0.1 W
Output power (maximum)	1 W r.m.s. at 6 V
Voltage gain (1 kHz)	39 dB
Printed circuit board size approximately	32 mm x 32 mm

Table 1.5   Specification of prototype TDA7052 1 W amplifier

TDA7052 1 watt power amplifier parts list

Resistors — All 1% 0.6 W metal film

| R1 | 1 k | 1 | (M1K) |
| RV1 | 10 k log lin pot | 1 | (JM77J) |

Capacitors

C1	2μ2F 63 V min electrolytic	1	(YY32K)
C2	100 nF ceramic	1	(YR75S)
C3	100 μF 16 V min electrolytic	1	(RA55K)

Semiconductors

| IC1 | TDA7052 | 1 | (UK79L) |

Miscellaneous

P1–11	1 mm PCB pins	1 Pkt	(FL24B)
	DIL socket 8-pin	1	(BL17T)
	printed circuit board	1	(GE41U)
	constructors' guide	1	(XH79L)

TDA1514A power amplifier

The TDA1514A is a high quality power amplifier integrated circuit which is compatible with a wide range of source material, including compact disc.

IC description

The TDA1514A is supplied in a 9-pin single in-line (SIL) plastic package (in the SOT131A format). The amplifier integrated circuit is used with a symmetrical power supply, with or without the bootstrap. The device features protection against a.c. and d.c. short circuits when used with symmetrical supplies. It also includes an output mute circuit preventing *clicks* and *pops* during switch on and switch off, eliminating the possibility of damage to delicate speakers. The amplifier is also protected against thermal runaway and includes SOAR (safe operating area region) protection making the device almost indestructible. An internal block diagram of the device is shown in Figure 1.26, and the pinout details are shown in Figure 1.27. Graph 1.1 shows the SOAR protection characteristics and Graph 1.2 shows the thermal derating curve.

Circuit description

The amplifier uses the circuit shown in Figure 1.28. The circuit also includes the previously mentioned bootstrap.

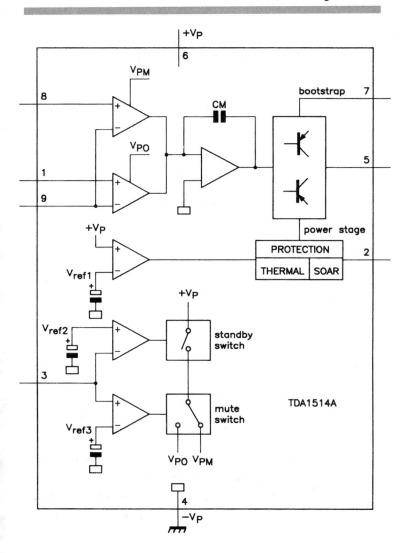

Figure 1.26   Internal block diagram

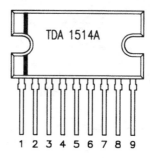

1. Non—inverting Input (#1)
2. SOAR
3. MUTE
4. —VE
5. Output
6. +VE
7. Bootstrap
8. 0V Non—inverting Input (#2)
9. Inverting Input

Figure 1.27   IC pinout

If the circuit is used without the bootstrap, pin 7 must be connected to pin 6 and the associated components (resistors R4 and R5, and capacitors C5 and C6) removed; the power output will be reduced by approximately 4 watts.

The input impedance of the amplifier module is 20 k determined by resistor R1. The gain of the amplifier is adjustable over the range 20 dB to 46 dB, determined by resistors R2 and R3. See Table 1.7 to select the required input sensitivity.

SOAR
protection
curve

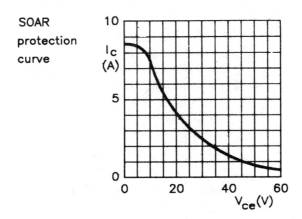

Graph 1.1   SOAR protection

Power
derating
curve

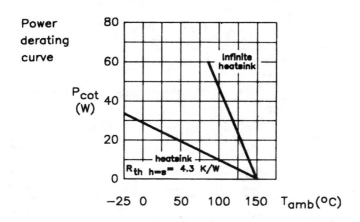

Graph 1.2   Thermal derating

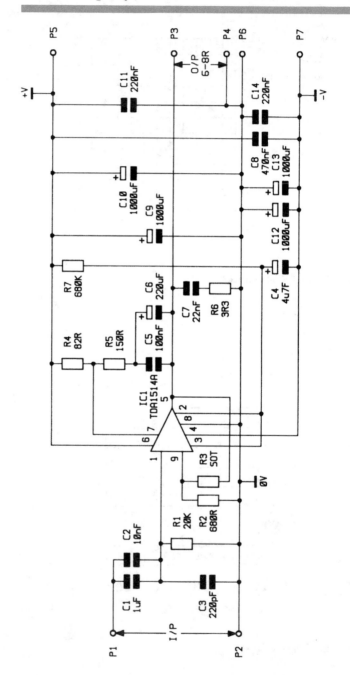

Figure 1.28 Application circuit

Power supply requirements

The integrated circuit has a wide supply range, the minimum requirements being ±9 V to a maximum of ±30 V. To deliver a 50 W into a 4 Ω the amplifier requires a 23 V supply and to deliver 40 W into 8 Ω the amplifier requires a 27.5 V supply; see Table 1.6.

Printed circuit board

A high quality glass fibre printed circuit board, with a silk screened component legend to aid construction is available. The printed circuit board has a 0 V star earth to minimise current return interaction from the speaker, amplifier and input, this will also help to maximise sound quality. Figure 1.29 shows the printed circuit board and component legend. Figure 1.30 shows the required leg shape and drilling details for mounting the TDA1514A integrated circuit onto a heatsink.

Test Conditions: V_p = ±27.5 V, R3 = 20 kΩ, input signal 1 kHz sinewave.

Rated output power	
R1 = 8 Ω	39 W r.m.s.
R1 = 4 Ω	78 W r.m.s.
Supply current (full output)	
R1 = 8 Ω	2.2 A
R1 = 4 Ω	4.4 A
Quiescent supply current	55.2 mA
THD at −3 dB of full output	0.1%
Signal-to-noise ratio	99 dB
Frequency response	20 Hz to 25 kHz −3 dB

Specifications of prototype

Audio IC projects

Parameter	Conditions	Symbol	Min	Typ	Max
Supply voltage range pin 6 and pin 4 wrt pin 8		V_p	±9 V		±30 V
Total quiescent current	$V_p = ±27.5$ V	I_{tot}		60 mA	
Output power (r.m.s.)	THD = −60 dB $V_p = ±27.5$ V $R_L = 8\ \Omega$	P_o		40 W	
	$V_p = ±23$ V $R_L = 4\ \Omega$	P_o		60 W	
Closed loop voltage gain	Determined externally	G_c		30 dB	
Input resistance	Determined externally	R_I		20 kΩ	
Signal plus noise to noise ratio	$P_o = 50$ mW	(S + N) ÷ N		82 dB	
Supply voltage ripple rejection	$f = 100$ Hz	SVRR		72 dB	

wrt = with respect to.

Table 1.6   Quick reference data

Input sensitivity	R3 value
300 mV	36 kΩ
500 mV	20 kΩ
1 V	10 kΩ
2 V	5k1Ω

Table 1.7   Input sensitivity

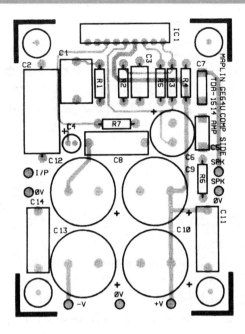

Figure 1.29   PCB component legend

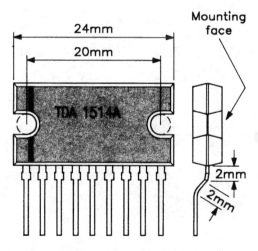

Figure 1.30   IC leg shaping, heatsink mounting

Audio IC projects

Wiring details

Figure 1.31 shows the module with the minimum power supply required, which consists of a transformer and bridge rectifier. Figure 1.32 shows the module wired up to an existing supply, note the speaker 0 V returns to the main power supply.

Power supply

The amplifier requires a supply of ±27.5 V, preferably smoothed. The absolute maximum voltage being ±30 V, which should be avoided if driving a load of less than 8 Ω. The supply must be capable of delivering peak currents of 8 amps or more, otherwise distortion will occur at full volume during loud passages and bass transients.

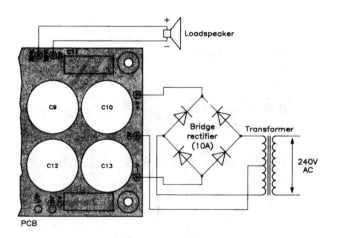

Figure 1.31   Minimum supply requirements

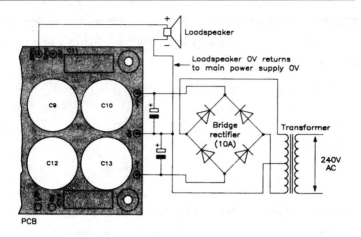

Figure 1.32   Using an existing supply

Thermal rating

Note: thermal resistance is shown written $K.W^{-1} = K/W$ (kelvin per watt).

The theoretical maximum power of dissipation for an output power of 40 watts is:

$$V_P^2 \div (2 \times \pi^2 \times R_L) = 19\,W$$

where $V_P = \pm27.5\,V$; $R_L = 8\,\Omega$.

Thermal resistance from junction to base: $R_{th\ j-mb} = 1\,K.W^{-1}$

41

With an ambient temperature of 50°C and a maximum junction temperature of 150°C, the total thermal resistance is:

$$R_{th-j} = (150 - 50) \div 19 = 5.3 \text{K.W}^{-1}$$

Taking into account power dissipation of the package (1 K^{-w}):

$$5.3 - 1 = 4.3 \text{K.W}^{-1}$$

A heatsink of thermal resistance 4.3 k.W^{-1} or less is required.

Applications

The TDA1514A amplifier kit is designed to provide a small, low cost hi-fi equipment to be easily built. Applications include active speakers; achieved by removing the crossover unit from the speaker enclosure and powering each speaker directly from an amplifier. A suitable electronic crossover is then fitted between the pre-amplifier and power amplifiers.

Other applications for the TDA1514A module range from building a small *midi* amplifier to replacing blown power amplifier integrated circuits that are difficult to obtain, by simply picking up the signal at the preamplifier output, and using the existing supply.

The amplifier module is also ideal for use in high power car stereo applications, a switch mode power supply would be required to step up the 12 V car supply to ±25 V.

A complete kit of parts with four different value resistors for R3 is available, kit number LP43W. The amplifier will require a heatsink (not included in the kit) with a thermal rating of 4.3 K.W^{-1} or less; heatsink type 2E (2.1 K.W^{-1}) is ideal, the order code is HQ70M. An insulating washer must be used between the integrated circuit and heatsink if the heatsink is electrically connected to earth, 0 V or any supply line, a suitable washer is available the order code is UL74R. The TDA1514A should be secured using either M3 bolts or No. 4 self-tapping screws of suitable length. Care should be exercised to avoid over tightening fixings, which will result in damage to the plastic package.

Notes to characteristics

1   measured with two superimposed signals of 50 Hz and 7 kHz with an amplitude ratio of 4:1,

2   the closed loop gain is determined by the resistors R2 and R3, the gain is adjustable between 20 and 40 dB,

3   the input impedance of the circuit is determined by the bias resistor R1,

4   the noise output voltage is measured in a bandwidth of 20 Hz to 20 KHz with a source resistance of 2 kΩ,

5   the quiescent current into pin 2 determines the minimum supply voltage at which the mute function remains in operation, $V_p - V_n = I_{2\,tot} \times R4 + V_{m\,(on)\,max}$

Audio IC projects

Parameter	Conditions	Sym	Min	Typ	Max
Supply voltage range pin 6 and pin 4 wrt pin 8		V_p	±9 V	±30 V	
Maximum output current (peak value)		I_{OM} max	6.4 A		
Operating state					
Input voltage pin 3 to pin 4		V_{3-4}	6 V	7 V	
Total quiescent current	R1 = inf	I_{tot}	30 mA	60 mA	90 mA
Output power	THD = −60 dB	P_o	37 W	40 W	
	THD = −20 dB	P_o		51 W	
Output power	V_p = ±23 V THD = −60 dB				
	R1 = 8 Ω	P_o		28 W	
	R1 = 4 Ω	P_o		50 W	
Total harmonic distortion	P_o = 32 W	THD		−90 dB	−80 dB
Intermodulation distortion	P_o = 32 W note 1	d_{im}		80 dB	
Power bandwith	−3 dB THD = −60 dB Hz	B		20Hz–2kHz	
Slew rate		dV/dt		10 V.µs[1]	
Closed loop voltage gain	note 2	G_c		30 dB	
Open loop voltage gain		G_o		85 dB	
Input impedance	note 3	Z_i	1 MΩ		
Output impedance		Z_o			0.1 Ω
Signal-to-noise ratio	note 4 P_o = 50 mW	S/N	80 dB		
Output offset voltage		V_o		2 mV	
Input bias current		I_i		−0.1 µA	

Table 1.8   IC data

Parameter	Conditions	Sym	Min	Typ	Max
Mute state					
Voltage on					
pin 3		V_{34}	2 V		4.5 V
Output voltage	$V_{I(r.m.s.)} = 2\,V$				
	$f = 1\,kHz$	V_o		100 μV	
Ripple rejection		RR		70 dB	
Standby state					
Voltage on					
pin 3		V_{34}	0 V		1 V
Total quiescent					
current		I_{tot}		20 mA	
Ripple rejection		RR		70 dB	
Supply voltage					
to obtain					
standby state		$\pm V_P$	4.5 V		7.0 V

Table 1.8   IC data (continued)

Parameter	Symbol	Minimum	Maximum
Supply voltage			
pin 6 and pin 4 wrt pin 8	V_P		±30 V
Bootstrap voltage			
pin 7 wrt pin 4	V_{bstr}		70 V
Output current			
repetitive peak	I_0		8 A
Storage temp	T_{stg}	−65°C	+150°C
Thermal shut-down			
protection time	t_{pr}		1 hour
Short circuit			
protection time	t_{sc}		10 minutes
Mute voltage			
pin 3 wrt pin 4	V_m		7 V

wrt = with respect to.

Table 1.9   Maximum ratings

TDA1514 power amplifier parts list

Resistors — All 0.6 W 1% metal film

R1	20 k	1	(M20K)
R2	680 Ω	1	(M680R)
R3	36 k (see text)	1	(M36K)
R3	20 k (see text)	1	(M20K)
R3	10 k (see text)	1	(M10K)
R3	5k1 (see text)	1	(M5K1)
R4	82 Ω	1	(M82R)
R5	150 Ω	1	(M150R)
R6	3Ω3	1	(M3R3)
R7	680 k	1	(M680K)

Capacitors

C1	1 µF polylayer	1	(WW53H)
C2	10 nF polystyrene	1	(BX92A)
C3	220 pF polystyrene	1	(BX30H)
C4	4µ7F 63 V PC electrolytic	1	(FF03D)
C5	100 nF polylayer	1	(WW41U)
C6	220 µF 63 V PC electrolytic	1	(FF14Q)
C7	22 nF polylayer	1	(WW33L)
C8	470 nF polyester	1	(BX80B)
C9,10,12,13	1000 µF 50 V SMPS	4	(JL57M)
C11,14	220 nF polyester	2	(BX78K)

Semiconductor

IC1	TDA1514A-N7	1	(UK75S)

Miscellaneous

P1–7	1 mm PCB pins	1	(FL24B)
	printed circuit board	1	(GE64U)
	constructors' guide	1	(XH79L)
Optional	insulator T0218	1	(UL74R)
	heatsink type 2E	1	(HQ70M)
	self-tap screw No. $4 \times {}^3/_8$ in	1	(BF65V)

2 Pre-amplifiers and filters

Featuring:

LM13700 dual transconductance op-amp — part 1

The LM13700 is a dual transconductance operational amplifier featuring low distortion and a wide dynamic range. The package contains two transconductance op-amps which share a common power supply input but are otherwise completely separate. Two high impedance buffers, designed to suit the dynamic range of the op-amp are included in the package. Figure 2.1 shows the integrated circuit pinout while Table 2.1 shows the electrical characteristics of the device.

Voltage controlled filter

A transconductance op-amp such as the LM13700 can form the basis of a versatile voltage controlled filter (VCF). The LM13700 is particularly suitable for use in a voltage controlled filter circuit as the required buffers are included in the integrated circuit. Figure 2.2 shows the circuit of a basic voltage controlled low pass filter; this circuit acts as a unity gain buffer below the cut-off frequency. Above the cut-off frequency the circuit provides a roll off of 6 dB per octave. Higher order filters may be produced using additional amplifiers as illustrated by the two pole butterworth filter shown in Figure 2.3. The circuit diagram of a typical state variable filter is shown in Figure 2.4; this provides both a low pass and a bandpass characteristic with a roll off of approximately 6 dB per octave.

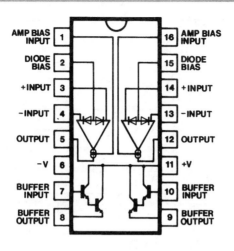

Figure 2.1   IC pinout diagram

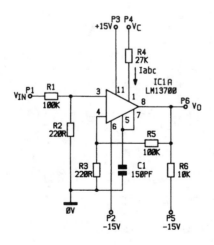

Figure 2.2   A typical voltage controlled low pass filter

Audio IC projects

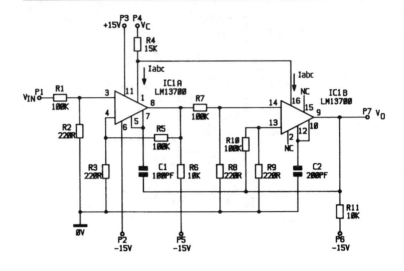

Figure 2.3   Example of a two pole butterworth filter

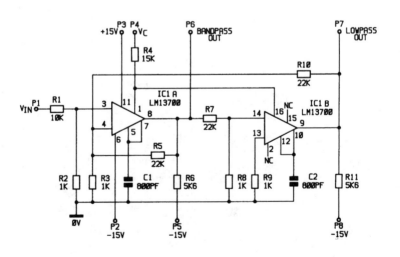

Figure 2.4   A typical voltage controlled state variable filter

Parameters	Operating conditions	Minimum	Typical	Maximum
Power suppy voltage		±2 V		±15 V
Differential input voltage				±5 V
Input offset voltage Iabc	5 μA	0.4 mV	4 mV	
Forward transconductance (gm)		6700 μS	9600 μS	13000 μS
gm tracking			0.3 dB	
Peak output voltage:				
Unloaded, Iabc	5 μA–500 μA			
Positive		+12 V	+14.2 V	
Negative		–12 V	–14.4 V	
Common mode range		±12 V	±13.5 V	
input resistance		10 kΩ	26 kΩ	
Slew rate	Unity gain compensated		50 V/μS	

Table 2.1   Electrical characteristics of LM13700 transconductance op-amp — specifications based on supply voltage ±15 V, amplifier bias current (Iabc) 500 μA, operating temperature 25°C and pins 2 and 5 open circuit (unless specified).

Voltage controlled oscillator

Effective voltage controlled oscillators can be realised using the LM13700. Figure 2.5 shows an example of a voltage controlled oscillator circuit that will produce both square and triangle wave outputs. The frequency of the oscillator is determined by the amplifier bias current (I_{ABC}) which may be varied between around 10 nA to 1 mA; with the component values shown in Figure 2.5, the oscillator will operate between approximately 2 Hz and

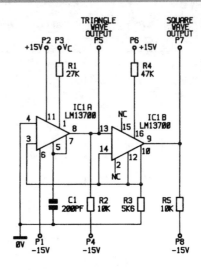

Figure 2.5 Example of a voltage controlled triangle/squarewave oscillator

200 kHz. The output amplitude of the oscillator is effectively determined by the value of resistor R5 and the magnitude of the amplifier output current. It is important that the differential input voltage does not exceed ±5 V; if this voltage is exceeded, the inputs will zener, resulting in distortion of the waveform.

Peak detect and hold

Figure 2.6 shows the diagram of a typical peak detector and hold circuit using the LM13700; this circuit detects the peak voltage of the signal and holds this value for a length of time determined by the value of capacitor C1.

Pre-amplifiers and filters

The circuit uses IC1(b) to turn on IC1(a) whenever input voltage (V_{in}) becomes more positive than output voltage (V_o). Pulling the output of IC1(b) low via diode D1 prevents capacitor C1 charging any more and therefore effectively holds the output voltage constant until C1 discharges.

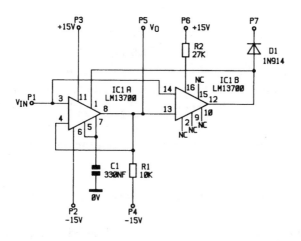

Figure 2.6   A typical peak detector and hold circuit

IC power supply requirements

The LM13700 integrated circuit requires a split rail power supply and will operate over a wide range of voltages between ±2 V and ±15 V. It is important that the supply is adequately decoupled in order to prevent the introduction of mains derived noise onto the supply rails. Decoupling close to the integrated circuit is also important as this helps to prevent any instability that may occur. For optimum performance a regulated power supply should be used. The integrated circuit current consumption is very much dependant on the application in which the device is being used but is generally no more than a few milliamps.

Printed circuit board

A high quality printed circuit board with printed legend is available to aid the construction of three simple circuits using the LM13700. The circuits that may be built using the printed circuit board are a voltage controlled state variable filter, a voltage controlled triangle/squarewave oscillator, and a peak detector and hold circuit.

Any one of the three circuits may be constructed using the same printed circuit board. A combined circuit diagram of the three applications is shown in Figure 2.7 for reference purposes; this is the circuit that was used to design the printed circuit board, the track layout of which is shown in Figure 2.8.

Figure 2.9 shows the circuit diagram of the voltage controlled state variable filter; this circuit operates from a single rail supply and may be powered from a wide range of voltages between 4 V and 30 V (maximum). Power supply connections are made to P1(+V) and P3(0 V). An input signal is applied between P4 and P12 and the filter produces a bandpass output on P6, with a low pass output on P8. A control voltage (30 V maximum) applied between P5 and 0 V determines the operating frequency of the filter (see Table 2.2).

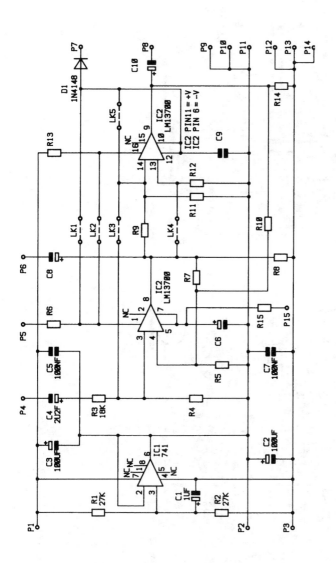

Figure 2.7 Combined circuit to which the PCB is designed

55

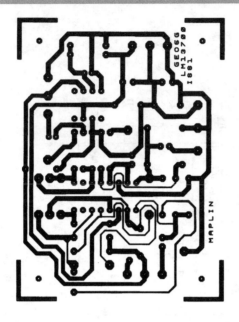

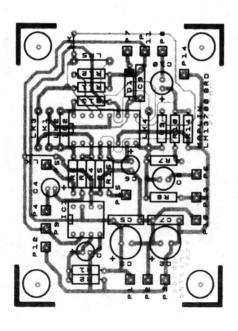

Figure 2.8   PCB legend and track

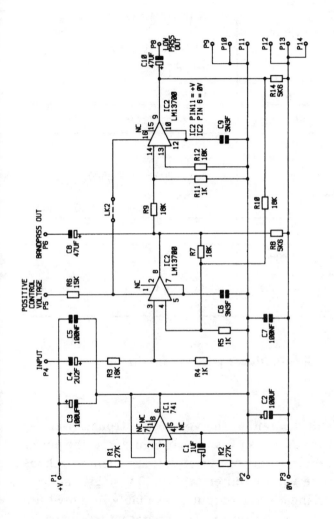

Figure 2.9 Circuit to build the VCF

57

Audio IC projects

Voltage controlled state variable filter:

Power supply voltage	30 V d.c.
Current drain	10 mA
Bandpass filter centre frequency	20 Hz–20 kHz (variable)
Low pass filter cut-off frequency	20 Hz–20 kHz (variable)
Roll off	6 dB per octave
Load impedance	10 kΩ
Maximum input voltage	1 V r.m.s.

Voltage controlled triangle/squarewave oscillator:

Power supply voltage	30 V
Current drain	10 mA
Output frequency	20 Hz–20 kHz (variable)
Load impedance	10 kΩ
Output voltage	5 V peak to peak (with 30 V supply)

Peak/hold circuit:

Power supply voltage	±15 V d.c.
Current drain	8 mA
Maximum input voltage	8 V peak to peak
Load impedance	1 kΩ

Table 2.2　Specification of prototype circuits

The circuit diagram of the voltage controlled triangle/squarewave oscillator is shown in Figure 2.10; the power supply requirements of this circuit are similar to those of the state variable filter (4 V to 30 V single rail supply). A triangle wave output is available on P6 with a squarewave output on P8. The operating frequency of the oscillator is controlled by a voltage (30 V maximum) applied between P5 and 0 V.

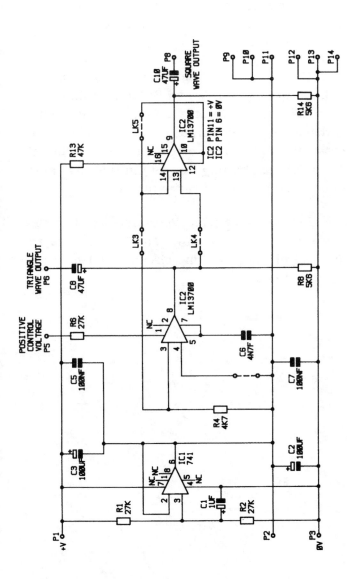

Figure 2.10 Circuit to build the oscillator

59

Audio IC projects

Figure 2.11 shows the diagram of the peak/hold circuit; this circuit operates from a ±4 V to ±15 V, split rail supply only. Power supply connections are made to P1(+V), P2(0 V) and P3(–V).

The input signal is applied between P4 and P9 and the output is taken between P6 and P10. Connecting P7 to P14 holds the output voltage level constant until capacitor C6 discharges. The output can be reset to 0 V by connecting P15 to 0 V. For some applications a directly coupled input may be required and in this case a link should be fitted in place of capacitor C4.

Figure 2.12 shows some typical performance characteristics for the LM13700 and Table 2.2 shows the specification of the prototype circuits built using the printed circuit board.

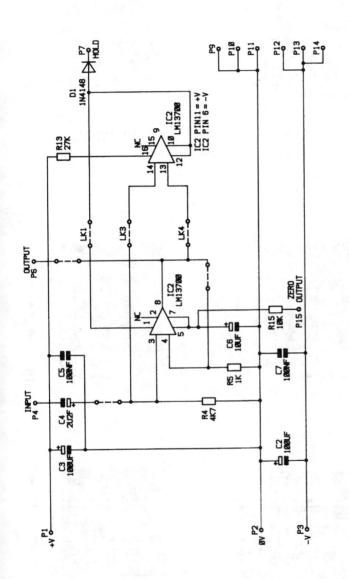

Figure 2.11 Circuit to build the peak and hold

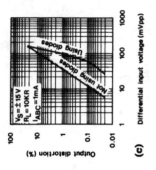

(a)

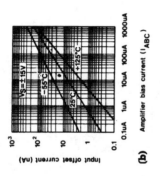

(b)

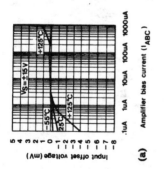

(c)

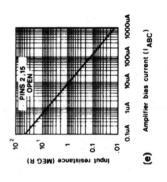

(d)

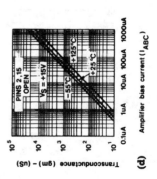

(e)

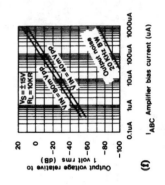

(f)

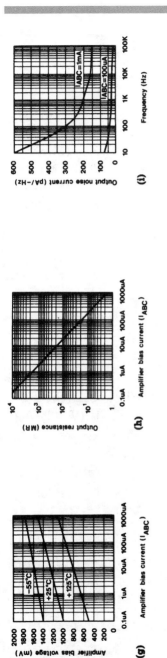

Figure 2.12 (a) input offset voltage (b) input offset current (c) distortion vs differential input voltage (d) transconductance vs amplifier bias current (e) input resistance vs amplifier bias current (f) output voltage vs amplifier bias current (g) amplifier bias voltage vs amplifier bias current (h) output resistance vs amplifier bias current (i) output noise current vs frequency

Audio IC projects

Voltage controlled filter parts list

Resistors — All 1% 0.6 W metal film

R1,2	27 k	2	(M27K)
R3,7,9, 10,12	18 k	5	(M18K)
R4,5,11	1 k	3	(M1K)
R6	15 k	1	(M15K)
R8,14	5k6	2	(M5K6)
R13	not fitted		
R15	not fitted		

Capacitors

C1	1 µF 100 V PC electrolytic	1	(FF01B)
C2,3	100 µF 35 V PC electrolytic	2	(JL19V)
C4	2µ2F 100 V PC electrolytic	1	(FF02C)
C5,7	100 nF polylayer	2	(WW41U)
C6	3n3F ceramic	1	(WX74R)
C8,10	47 µF 50 V PC electrolytic	2	(JL16S)
C9	3n3F polylayer	1	(WW25C)

Semiconductors

IC1	LM741CN	1	(QL22Y)
IC2	LM13700	1	(YH64U)
D1	not fitted		

Links

LK1	not fitted
LK2	fitted
LK3	not fitted
LK4	not fitted
LK5	not fitted

Miscellaneous

P1–6,8–14	1 mm PCB pins	1 (FL24B)
	DIL socket 8-Pin	1 (BL17T)
	DIL socket 16-Pin	1 (BL19V)
	printed circuit board	1 (GE06G)
	constructors guide	1 (XH79L)

Voltage controlled oscillator parts list

Resistors — All 1% 0.6 W metal film

R1,2,6	27 k	3	(M27K)
R3	not fitted		
R4	4k7	1	(M4K7)
R5	fit link		
R7	not fitted		
R8,14	5k6	2	(M5K6)
R9	not fitted		
R10	not fitted		
R11	not fitted		
R12	not fitted		
R13	47 k	1	(M47K)
R15	not fitted		

Capacitors

C1	1 µF 100 V PC electrolytic	1	(FF01B)
C2,3	100 µF 35 V PC electrolytic	2	(JL19V)
C4	not fitted		
C5,7	100 nF polyester	2	(WW41U)
C6	4n7F ceramic	1	(WX76H)
C8,10	47 µF 50 V PC electrolytic	2	(JL16S)
C9	not fitted		

Audio IC projects

Semiconductors

IC1	LM741CN	1	(QL22Y)
IC2	LM13700N	1	(YH64U)
D1	not fitted		

Links

LK1	not fitted
LK2	not fitted
LK3	fitted
LK4	fitted
LK5	fitted

Miscellaneous

P1–3,5,6, 8–14	1 mm PCB pins	1	(FL24B)
	DIL socket 8-pin	1	(BL17T)
	DIL socket 16-pin	1	(BL19V)
	printed circuit board	1	(GE06G)
	constructors guide	1	(XH79L)

Peak/hold circuit parts list

Resistors — All 1% 0.6 W metal film

R1	not fitted		
R2	not fitted		
R3	fit link		
R4	4k7	1	(M4K7)
R5	1 k	1	(M1K)
R6	not fitted		

R7	fit link		
R8	not fitted		
R9	not fitted		
R10	not fitted		
R11	not fitted		
R12	not fitted		
R13	27 k	1	(M27K)
R14	not fitted		
R15	10 k	1	(M10K)

Capacitors

C1	not fitted		
C2,3	100 µF 35 V PC electrolytic	2	(JL19V)
C4	2µ2F 100 V PC electrolytic	1	(FF02C)
C5,7	100 nF polyester	2	(WW41U)
C6	10 µF 50 V PC electrolytic	1	(FF04E)
C8	fit link		
C9	not fitted		
C10	not fitted		

Semiconductors

IC1	not fitted		
IC2	LM13700N	1	(YH64U)
D1	1N4148	1	(QL80B)

Links

LK1	fitted
LK2	not fitted
LK3	fitted
LK4	fitted
LK5	not fitted

Audio IC projects

Miscellaneous

P1–4,6,7,
9–15   1 mm PCB pins   1 (FL24B)

DIL socket 16-Pin   1 (BL19V)

printed circuit board   1 (GE06G)

constructors guide   1 (XH79L)

LM13700 dual transconductance op amp — part 2

The LM13700 is a dual transconductance operational amplifier incorporating linearising diodes to reduce distortion at high output levels. Two internal Darlington buffer transistors, designed to suit the dynamic range of the amplifiers are included in the package. Both amplifiers share the same supply connections, but are otherwise completely separate. Figure 2.13 shows the integrated circuit pinout while Table 2.3 shows the electrical characteristics of the device.

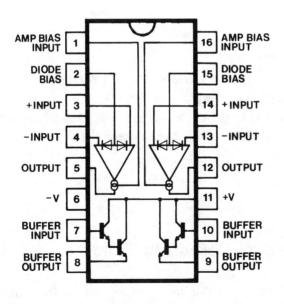

Figure 2.13 IC pinout

Audio IC projects

Parameters	Operating conditions	Min	Typ	Max
Power supply voltage		±2 V		±15 V
Differential input voltage				±5 V
Amplifier bias current (Iabc)				2 mA
Diode current (I_d)				2 mA
Input offset voltage	Iabc 5 µA	0.4 mV	4 mV	
Forward transconductance (gm)		6700 µS	9600 µS	13000 µS
gm tracking			0.3 dB	
Peak output voltage: unloaded, Iabc 5 µA–500 µA				
Positive		+12 V	+14.2 V	
Negative		–12 V	–14.4 V	
Common mode range		±12 V	±13.5 V	
Input resistance		10 kΩ	26 kΩ	
Slew rate	Unity gain compensated		50 V/µs	

Table 2.3   Electrical characteristics of LM13700N transconductance op-amp — specifications based on supply voltage ±15 V, amplifier bias current (Iabc) 500 µA, operating temperature 25°C and pins 2 and 15 open circuit (unless specified).

Voltage controlled amplifiers

Figure 2.14 shows the basic voltage controlled amplifier configuration. Preset resistor RV1 is adjusted to mini-mise the effect of the control signal on the output. The gain of the amplifier is determined by the control voltage (V_c). In order to achieve optimum signal-to-noise performance, amplifier bias current (I_{abc}), should be kept as high as possible. The linearising diodes help to prevent increased distortion at high output levels. It is recommended that the linearising diode current (I_d) is kept relatively high to enhance the effect of the diodes; a diode current of 1 mA is suitable for most purposes.

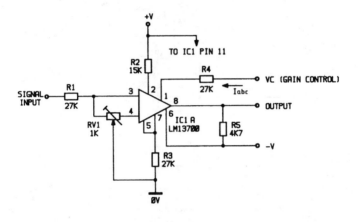

Figure 2.14 A typical voltage controlled amplifier (VCA)

The two amplifiers in the LM13700 package are very closely matched, making the device ideal for stereo volume control applications as illustrated by the circuit shown in Figure 2.15; this circuit typically provides a channel to channel gain tracking of less than 0.3 dB. Preset resistors RV1 and RV2 are adjusted to reduce the output offset voltage to a minimum and if the amplifier is capacitively coupled, these may be replaced by fixed resistors.

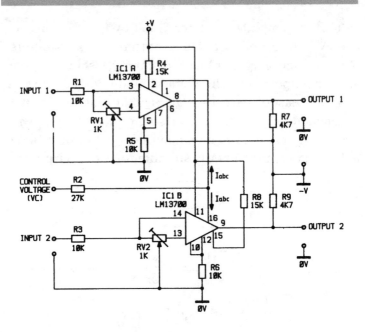

Figure 2.15   Stereo volume control

Automatic gain control

Figure 2.16 shows a typical example of an AGC amplifier using the LM13700; this circuit makes use of the fact that the gain of the amplifier can be controlled by varying the linearising diode current as well as the amplifier bias current. When the output voltage (V_o) reaches a high enough level to turn on the output buffer transistor, the rise in current through the linearising diodes reduces the gain of the amplifier so as to hold V_o at a constant level. The output amplitude level is adjusted using preset RV1.

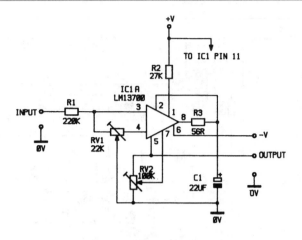

Figure 2.16   An example of an AGC amplifier

Pulse width modulation

A typical example of a pulse width modulator is shown
in Figure 2.17. A squarewave (clock) is applied at the
input and the circuit produces an output of variable pulse
width. The pulse width is determined by the control voltage (V_c) and the value of capacitor C2.

IC power supply requirements

The LM13700 integrated circuit is designed to be powered from a split rail supply and the device operates over
a wide range of voltages between ±2 V and ±15 V. For
optimum performance and to prevent instability, it is

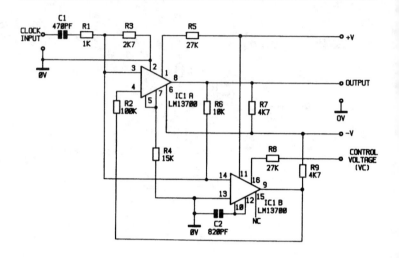

Figure 2.17   A typical pulse width modulator

important that adequate decoupling is used close to the integrated circuit. Amplifier current consumption depends on individual applications, but is generally no more than a few mA.

Printed circuit board

A high quality, fibreglass, multipurpose printed circuit board with printed legend is available for three simple applications of the LM13700; the circuits that can be built using the printed circuit board are:

Pre-amplifiers and filters

- a voltage controlled stereo amplifier,

- an AGC amplifier,

- a pulse width modulator.

Any one of the three circuits may be constructed on the same printed circuit board. Figure 2.18 shows the combined circuit diagram that was used to produce the printed circuit board and Figure 2.19 shows the component layout diagram. All three circuits have been designed using additional components (resistors R1 and R2, capcitor C1 and integrated circuit IC1) to allow operation from a single rail power supply. When being operated from a single rail supply, the input voltage may range from approximately 4 V to a maximum of 30 V.

Figure 2.20 shows the circuit diagram of the voltage controlled stereo amplifier. Power supply connections are made to P1 (+V) and P3 (0 V). Preset resistors RV1 and RV3 are included in the circuit to allow adjustment of output offset voltage and these should be adjusted for symmetrical clipping. As a general guide, for most purposes it is sufficient to set each preset to the centre of its travel. Input signals are applied to P4 (i/p), P5 (0 V) and P6 (i/p), P7 (0 V), with outputs being taken from P9 (o/p), P16 (0 V) and P10 (o/p), P17 (0 V), respectively. The gain of the amplifier is controlled by a positive voltage (maximum 30 V), applied to P8 (V_c) and P15 (0 V); if required, this voltage may be derived from the supply rail using a potentiometer (a suitable value is 10 kΩ). Resistor R14 provides the option of setting a minimum control voltage when a potentiometer is used and this component should be chosen to suit the application or

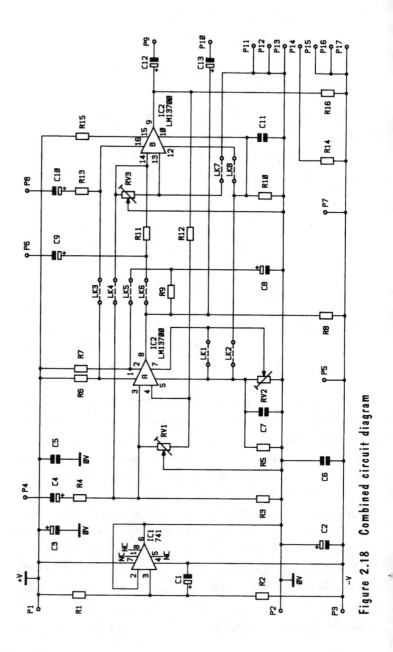

Figure 2.18 Combined circuit diagram

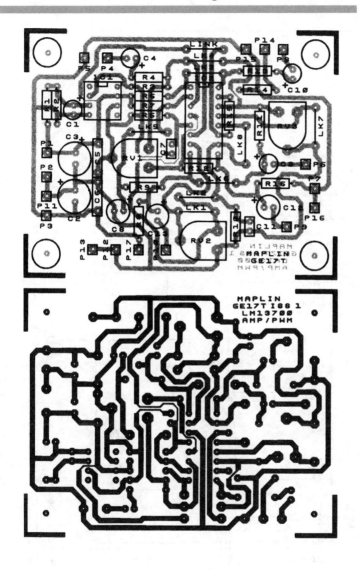

Figure 2.19   PCB layout diagram

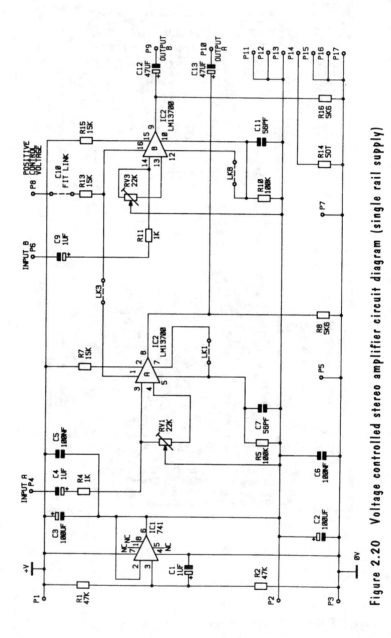

Figure 2.20 Voltage controlled stereo amplifier circuit diagram (single rail supply)

omitted as appropriate. If resistor R14 *is* used, the anti-clockwise end of the potentiometer track should be connected to P14 instead of P15.

Figure 2.21 shows the circuit diagram of the AGC amplifier. Supply connections are made to P1 (+V) and P3 (0 V). Input signals are applied to P4 (i/p) and P5 (0 V) and the output is taken from P9 (o/p) and P16 (0 V). Preset resistor RV1 should be adjusted for symmetrical clipping and as a general guide this preset should be set as accurately as possible to the centre of its travel. The amplitude at which the AGC action occurs is determined by the setting of preset resistor RV2.

Figure 2.22 shows the circuit diagram of the pulse width modulator. Power supply connections are made to P1 (+V) and P3 (0 V). For a squarewave input applied to P4 (i/p) and P5 (0 V), the circuit produces an output of variable pulse width on P10 (o/p) and P17 (0 V). The pulse width is determined by a control voltage applied to P8 (V_c) and P15 (0 V). Although the control voltage is usually directly coupled for most purposes, provision for capacitive coupling is provided by capacitor C10; this capacitor is therefore normally linked out. The pulse output is capacitively coupled as the integrated circuit's output does not swing all the way to 0 V; alternative methods of coupling may be used to suit individual applications.

Split rail supply

Although the circuits are designed to be used with a single rail power supply, all three circuits can be modified for operation from a ±2 V to ±15 V split supply; the nec-

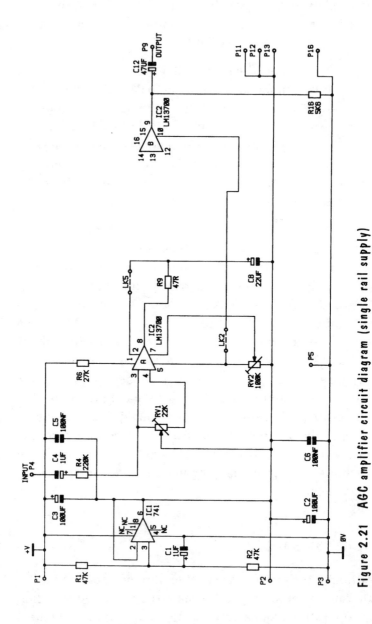

Figure 2.21 AGC amplifier circuit diagram (single rail supply)

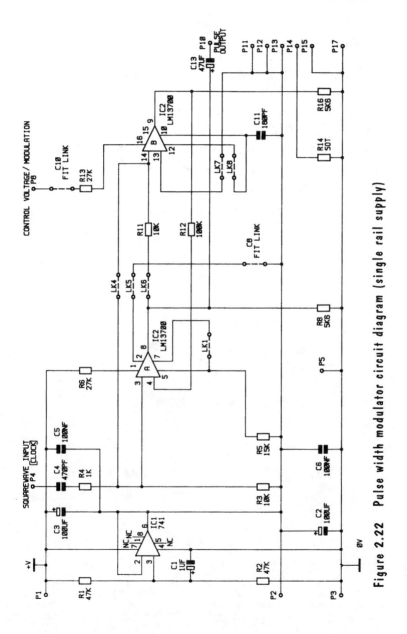

Figure 2.22 Pulse width modulator circuit diagram (single rail supply)

essary component changes are shown in Table 2.4. Capacitive output coupling may be required for the AGC amplifier depending on the application.

Note: for split rail operation, *power supply* connections are made to P1 (+V), P2 (0 V) and P3 (–V) and *signal* 0 V connections are made to P11, P12 and P13 (instead of P15, P16 and P17). Figure 2.23 shows some typical performance characteristics for the LM13700 and Table 2.5 shows the specification of the prototype circuits built using the printed circuit board.

Circuit	Do not fit these components	Fit links in place of these components
Voltage controlled stereo amplifier	R1,R2,C1,IC1	C12,C13
AGC amplifier	R1,R2,C1,IC1	C12
Pulse width modulator	R1,R2,C1,IC1	C13

Table 2.4 Circuit modifications for split supply operation

Voltage controlled stereo amplifier

Power supply voltage	4 V to 30 V d.c.
Power supply current	18 mA at 30 V
Voltage gain (control voltage = 30 V, supply voltage = 30 V)	25 (28 dB)
Bandwidth (−3 dB)	20 Hz to 25 kHz
Output load	10 kΩ

AGC amplifier

Power supply voltage	4 V to 30 V d.c.
Power supply current	10 mA at 30 V
Output load	10 kΩ

Pulse width modulator

Power supply voltage	4 V to 30 V d.c.
Power supply current	14 mA at 30 V
Output load	10 kΩ
Frequency range	50 Hz to 40 kHz
Maximum control voltage	30 V

Table 2.5 Specification of prototype circuits built using the PCB

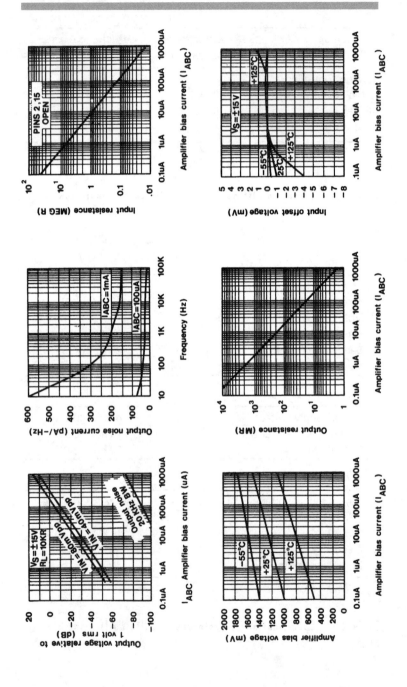

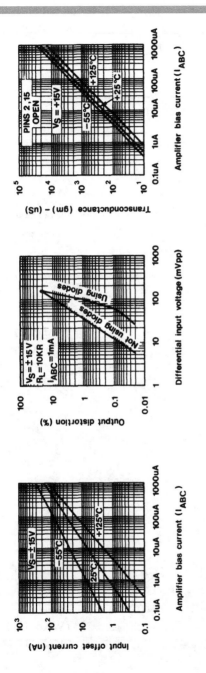

Figure 2.23 Typical performance characteristics for the LM13700

Pulse width modulator parts list

Resistors — All 1% 0.6 W metal film

R1	47 k	1	(M47K)
R2	47 k	1	(M47K)
R3	10 k	1	(M10K)
R4	1 k	1	(M1K)
R5	15 k	1	(M15K)
R6	27 k	1	(M27K)
R7	not fitted		
R8	5k6	1	(M5K6)
R9	not fitted		
R10	not fitted		
R11	10 k	1	(M10K)
R12	100 k	1	(M100K)
R13	27 k	1	(M27K)
R14	SOT		
RI5	not fitted		
R16	5k6	1	(M5K6)
RV1	not fitted		
RV3	not fitted		
RV3	not fitted		

Capacitors

C1	1 µF 100 V PC electrolytic	1	(FF01B)
C2	100 µF 35 V PC electrolytic	1	(JL19V)
C3	100 µF 35 V PC electrolytic	1	(JL19V)
C4	470 pF ceramic	1	(WX64U)
C5	100 nF polylayer	1	(WW41U)

C6	100 nF polylayer	1	(WW41U)
C7	not fitted		
C8	fit link		
C9	not fitted		
C10	fit link		
C11	180 pF ceramic	1	(WX59P)
C12	not fitted		
C13	47 µF 50 V PC electrolytic	1	(JL16S)

Semiconductors

IC1	LM741CN	1	(QL22Y)
IC2	LM13700N	1	(YH64U)

Links

LK1	fitted
LK2	not fitted
LK3	not fitted
LK4	fitted
LK5	fitted
LK6	fitted
LK7	fitted
LK8	fitted

Miscellaneous

1 mm PCB pins	1	(FL24B)
DIL socket 8-Pin	1	(BL17T)
DIL socket 16-Pin	1	(BL19V)
printed circuit board	1	(GE17T)
constructors guide	1	(XH79L)

Stereo VCA parts list

Resistors — All 1% 0.6 W metal film.

R1	47 k	1	(M47K)
R2	47 k	1	(M47K)
R3	not fitted		
R4	1 k	1	(M1K)
R5	100 k	1	(M100K)
R6	not fitted		
R7	15 k	1	(M15K)
R8	5k6	1	(M5K6)
R9	not fitted		
R10	100 k	1	(M100K)
R11	1 k	1	(M1K)
R12	not fitted		
R13	15 k	1	(M15K)
R14	SOT		
R15	15 k	1	(Ml5K)
R16	5k6	1	(M5K6)
RV1	22 k hor encl preset	1	(UH04E)
RV2	not fitted		
RV3	22 k hor encl preset	1	(UH04E)

Capacitors

C1	1 µF 100 V PC electrolytic	1	(FF01B)
C2	100 µF 35 V PC electrolylic	1	(JL19V)
C3	100 µF 35 V PC electrolytic	1	(JL19V)
C4	1 µF 100 V PC electrolytic	1	(FF01B)
C5	100 nF polylayer	1	(WW4lU)

C6	100 nF polylayer	1	(WW41U)
C7	56 pF ceramic	1	(WX53H)
C8	not fitted		
C9	1 µF 100 V PC electrolytic	1	(FF01B)
C10	fit link		
C11	56 pF ceramic	1	(WX53H)
C12	47 µF 50 V PC electrolytic	1	(JL16S)
C13	47 µF 50 V PC electrolyic	1	(JL16S)

Semiconductors

IC1	LM741CN	1	(QL22Y)
1C2	LM13700N	1	(YH64U)

Links

LK1	fitted
LK2	not fitted
LK3	fitted
LK4	not fitted
LK5	not fitted
LK6	not fitted
LK7	not fitted
LK8	fitted

Miscellaneous

1 mm PCB pins	1	(FL24B)
DIL socket 8-Pin	1	(BL17T)
DIL socket 16-Pin	1	(BL19V)
printed circuit board	1	(GE17T)
constructors guide	1	(XH79L)

AGC amp parts list

Resistors — All 1% 0.6 W metal film

R1	47 k	1	(M47K)
R2	47 k	1	(M47K)
R3	not fitted		
R4	220 k	1	(M220K)
R5	not fitted		
R6	27 k	1	(M27K)
R7	not fitted		
R8	not fitted		
R9	47 Ω	1	(M47R)
R10	not fitted		
R11	not fitted		
R12	not fitted		
R13	not fitted		
R14	not fitted		
R15	not fitted		
R16	5k6	1	(M5K6)
RV1	22 k hor encl preset	1	(UH04E)
RV2	100 k hor encl preset	1	(UH06G)
RV3	not fitted		

Capacitors

C1	1 µF 100 V PC electrolytic	1	(FF01B)
C2	100 µF 35 V PC electrolytic	1	(JL19V)
C3	100 µF 35 V PC electrolytic	1	(JL19V)
C4	1 µF 100 V PC electrolytic	1	(FF01B)
C5	100 nF polylayer	1	(WW41U)

C6	100 nF polylayer	1	(WW41U)
C7	not fitted		
C8	22 µF 50 V PC electrolytic	1	(JL12N)
C9	not fitted		
C10	not fitted		
C11	not fitted		
C12	47 µF 50 V PC electrolytic	1	(JL16S)
C13	not fitted		

Semiconductors

IC1	LM741CN	1	(QL22Y)
IC2	LM13700N	1	(YH64U)

Links

LK1	not fitted
LK2	fitted
LK3	not fitted
LK4	not fitted
LK5	fitted
LK6	not fitted
LK7	not fitted
LK8	not fitted

Miscellaneous

1 mm PCB pins	1	(FL24B)
DIL socket 8-Pin	1	(BL17T)
DIL socket 16-Pin	1	(BL19V)
printed circuit board	1	(GE17T)
constructors' guide	1	(XH79L)

91

MF10 universal switched capacitor filter

The MF10 effectively contains two CMOS active filter building blocks which can be configured with the addition of external components, to produce different second order functions (12 dB/octave). It is possible to use either of the blocks individually or alternatively the two blocks may be cascaded to provide higher order functions. Figure 2.24 shows the integrated circuit pinout diagram and Table 2.6 lists some typical electrical characteristics for the MF10.

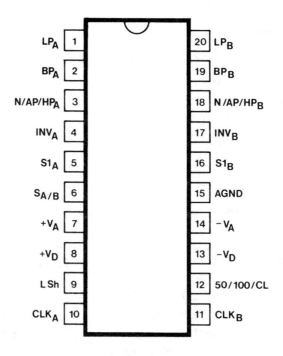

Figure 2.24 IC pinout (see opposite)

Pre-amplifiers and filters

Pin description

LP, BP, N/AP/HP   These are the lowpass, bandpass, notch or allpass or highpass outputs of each 2nd order section. The LP and BP outputs can sink typically 1 mA and source 3 mA. The N/AP/HP output can typically sink and source 1.5 mA and 3 mA, respectively.

INV   This is the inverting input of the summing op-amp of each filter. The pin has static discharge protection.

S1   S1 is a signal input pin used in the allpass filter configurations (see modes of operation 4 and 5). The pin should be driven with a source impedance of less than 1 kΩ.

$S_{A/B}$ It activates a switch connecting one of the inputs of the filter's 2nd summer either to analogue ground ($S_{A/B}$ low to $-V_A$) or to the lowpass output of the circuit ($S_{A/B}$ high to $+V_A$). This allows flexibility in the various modes of operation of the integrated circuit. $S_{A/B}$ is protected against static discharge.

$+V_A$, $+V_D$ Analogue positive supply and digital positive supply. These pins are internally connected through the IC substrate and therefore $+V_A$ and $+V_D$ should be derived from the same power supply source. They have been brought out separately so they can be bypassed by separate capacitors, if desired. They can be externally tied together and bypassed by a single capacitor.

$-V_A$, $-V_D$ Analogue and digital negative supply respectively. The same comments as for $+V_A$ and $+V_D$ apply here.

L Sh   Level shift pin: it accommodates various clock levels with dual or single supply operation. With dual ±5 V supplies, the MF10 can be driven with CMOS clock levels (±5 V) and the L Sh pin should also be tied either to the system ground or to the negative supply pin. If the same supplies as above are used but T²L clock levels, derived from 0 V to 5 V supply, are only available, the L Sh pin should be tied to the system ground. For single supply operation (0 V and 10 V) the $-V_D$, $-V_A$ pins should be connected to the system ground, the AGND pin should be biased at 5 V and the L Sh pin should also be tied to the system ground. This will accommodate both CMOS and T²L clock levels.

CLK(A or B)   Clock inputs for each switched capacitor filter building block. They should both be of the same level (T²L or CMOS). The level shift (L Sh) pin description discusses how to accommodate their levels. The duty cycle of the clock should preferably be close to 50% especially when clock frequencies above 200 kHz, are used. This allows the maximum time for the op-amps to settle which yields optimum filter operation.

50/100/CL   By tying the pin high a 50:1 clock to filter centre frequency operation is obtained. Tying the pin at mid supplies (i.e., analogue ground with dual supplies) allows the filter to operate at a 100:1 clock to centre frequency ratio. When the pin is tied low, a simple current limiting circuitry is triggered to limit the overall supply current down to about 2.5 mA. The filtering action is then aborted.

AGND   Analogue ground pin; it should be connected to the system ground for dual supply operation or biased at mid supply for single operation. The positive inputs of the filter op-amps are connected to the AGND pin so *clean* ground is mandatory. The AGND pin is protected against static discharge.

Audio IC projects

Parameter	Conditions	Min	Typ	Max
Supply voltage		±4 V	±5 V	
Absolute maximum				±7 V
Supply current			8 mA	10 mA
Maximum clock frequency		1 MHz	1.5 MHz	
Clock feedthrough			10 mV	
Crosstalk			50 dB	
Operating temperature	Absolute maximum	0°C		70°C
Storage temperature	Absolute maximum			150°C
Power dissipation	Absolute maximum			500 mW

Table 2.6   MF10 typical electrical characteristics

Integrated circuit description

The MF10 is supplied in a 20-pin dual-in-line package
which comprises two active filter blocks. Each block has
three separate output pins: pin 1(20) and pin 2(19) provide lowpass and bandpass outputs respectively, while
pin 3(18) may be configured to perform either notch,
highpass or allpass functions. The cut-off frequency of
the filter is partially determined by the frequency of the
clock and partially by external resistor ratios depending on the circuit configuration used. Both lowpass and
bandpass outputs can be configured so that the cut-off
frequency is a direct submultiple of the clock frequency.
The centre frequency of the notch and allpass circuits is
always directly dependent on clock frequency.

Integrated circuit power supply requirements

The MF10 requires a split rail power supply of typically ±5 V at around 8 mA. Separate pins are provided for analogue and digital supplies and separate supply rails should be used if minimal interaction between the analogue and digital sections of the circuit is to be achieved. Both the analogue and digital supplies should be decoupled individually; this is particularly important at high frequencies to prevent the clock from modulating the output.

DIY module

Figure 2.25 shows the circuit diagram of a general-purpose application module and Figure 2.26 shows a possible printed circuit legend using the MF10 universal switched capacitor filter integrated circuit.

The suggested module has an on-board clock and 6 separate buffered outputs (3 for each filter block). Facility is provided for voltage control of the clock frequency and selectable clock/centre frequency ratio.

The circuit caters for lowpass, bandpass and notch filter characteristics with the option for alternative highpass operation. If required the on-board clock may be disconnected to allow the use of an external clock oscillator.

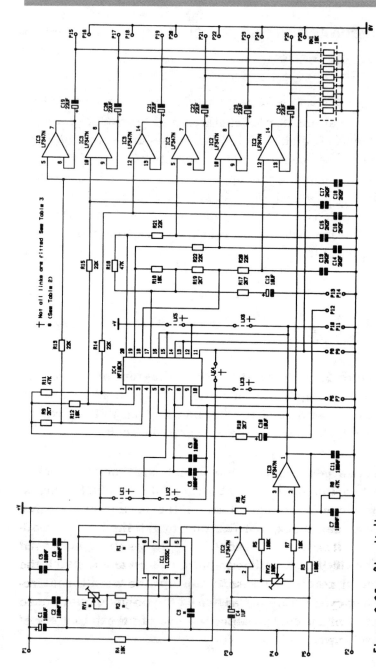

Figure 2.25 Circuit diagram

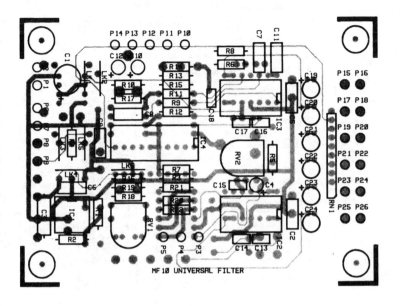

Figure 2.26   Legend of suggested printed circuit

Connection information is given in Figure 2.27. Inputs are connected between P11(0 V) and P12(i/p) and P13(i/p) and P14(0 V). Output signals are taken from P15–P26. The frequency of the internal clock is adjusted by preset resistor RV2, the range of which is set by resistors R1 and R2 and capacitor C3. The circuit has the basic component values to allow it to operate on any one of three different ranges and Table 2.7 shows the options available. It should be noted that the ranges shown are based on 100/1 clock/centre frequency ratio and apply to the basic notch/bandpass/lowpass mode only. Clock frequency may also be adjusted by applying a voltage to the voltage control input (P3–P5) and the sensitivity of this input is adjusted using preset resistor RV2.

Audio IC projects

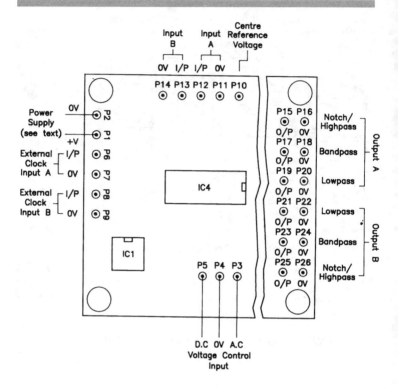

Figure 2.27   Wiring diagram of possible printed circuit module

Frequency range	R1	R2	RV1	C3
10 Hz–150 Hz	10 k	47 k	1 M	1000 pF
100 Hz–1.4 kHz	10 k	47 k	1 M	100 pF
1 kHz–8 kHz	1 k	4k7	100 k	100 pF

Table 2.7   Approximate frequency ranges and corresponding component values

Pre-amplifiers and filters

Some of the facilities provided by the circuit can be selected by fitting links in the printed circuit board. There are 6 links and Table 2.8 shows which of these must be fitted for each function. The links marked with a tick must be fitted when the function shown in the corresponding left hand column of the table is selected; those marked with an X symbol in the same column must be left out. In a typical example the module could be used to provide two sets of notch, bandpass and lowpass outputs using the on-board clock with a 100/1 clock/centre frequency ratio and in this case LK1, LK3, LK4 and LK6 only would be fitted.

If the internal clock oscillator is not required, then links LK3 and LK4 should not be fitted. An external clock signal may then be applied to P6(i/p) and P7(0 V) and P8(i/p) and P9(0 V). The maximum external clock frequency is typically 1.5 MHz. It should also be remembered that output filter capacitors C13–C18, together with resistors R13–R15 and R20–R22, limit the bandwidth, so the value of these components must therefore be taken into consideration when evaluating the maximum operating frequency of the circuit.

Function	Link number					
	1	2	3	4	5	6
Notch/lowpass/bandpass	✔	✗	-	-	-	-
Highpass/lowpass/bandpass	✗	✔	-	-	-	-
Internal clock (input A)	-	-	✔	-	-	-
Internal clock (input B)	-	-	-	✔	-	-
Clock/centre frequency ratio 50/1	-	-	-	-	✔	✗
Clock/centre frequency ratio 100/1	-	-	-	-	✗	✔

Table 2.8 Link function table

Audio IC projects

Power supply requirements

The circuit requires a power supply of between 8 V and 12 V which is capable of supplying at least 50 mA. For optimum performance, a regulated power supply should be used, which is properly smoothed to prevent the introduction of mains derived noise into the system. Separate high frequency decoupling for the analogue and digital supplies is provided for in the circuit design. Power supply connections are made to P1(+V) and P2(0 V).

Applications

The filter has many varied applications ranging from sound effect generation in electronic musical instruments to audio frequency filtering for communications equipment. The filter frequency may be swept by applying an alternating voltage to the voltage control input of the circuit, between P3 and P4. It should be noted that P3 is capacitively coupled and is therefore suitable for a.c. signals whereas P5 is directly coupled and more suitable for use as a d.c. control voltage input.

Typical frequency response characteristics, using the component values supplied in the kit are shown in Figure 2.28, Figure 2.29 and Figure 2.30. The responses shown are based on a centre frequency of 1 kHz and an input level of 770 mV. Variations in filter response may be obtained using different component values.

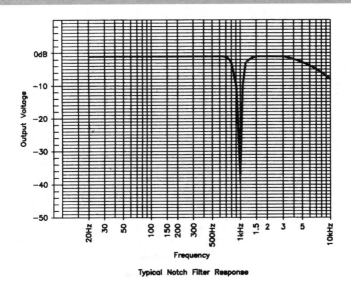

Figure 2.28   Typical notch filter response

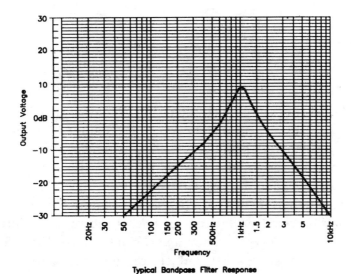

Figure 2.29   Typical bandpass filter response

101

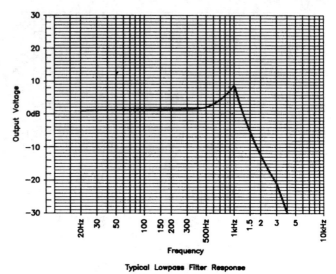

Typical Lowpass Filter Response

Figure 2.30   Typical lowpass filter response

If a sharper cut-off characteristic is required than is possible with one filter then two or more filter blocks can
be cascaded to produce a sharper response. If the same
clock is used for both blocks then the filter cut-off frequencies should track relatively closely. Depending on
the number of stages used, additional input attenuation
may be required in front of one or more of the filter
blocks to prevent overloading and in this case, the value
of input resistors R10 and R17 can be increased. Alternatively an external attenuation network may be used
for convenience.

Pre-amplifiers and filters

Because the MF10 is based on a digital system, the output waveform is made up of a series of steps corresponding to the clock frequency. Although the on-board filter network removes most of the digital content of the signal it only provides very basic filtering and a higher level of distortion is therefore noticeable at low frequencies. If the filter is going to be used over a limited range, additional low pass filtering may be used to smooth the signal.

Power supply voltage	8 V–12 V
Power supply current (quiescent)	29 mA at 10 V
Internal clock frequency range	See table 3
Clock/centre frequency ratio	Selectable 100/1 or 50/1 except highpass
Outputs	Notch, bandpass, lowpass with highpass option
Maximum voltage gain (tested at 1 kHz at peak of bandpass response)	10 dB
Maximum input voltage	0.9 V r.m.s.
Printed circuit board dimensions approximately	73 mm x 102 mm

Table 2.9   Specification of prototype MF10 filter module

MF10 universal switched capacitor filter parts list

Resistors — All 1% 0.6 W metal film

R1,2	see additional parts below		
R4,7, 12,18	10 k	4	(M10K)
R3,5	100 k	2	(M100K)
R6,8, 11,16	47 k	4	(M47K)
R9,10, 17,19	2k7	4	(M2K7)
R13,14, 15,20, 21,22	22 k	6	(M22K)
RV1	see additional parts below		
RV2	100 k hor encl preset	1	(UH06G)
RN1	10 k SIL array	1	(RA30H)

Capacitors

C1	100 µF 16 V min electrolytic	1	(RA55K)
C2,5–9, 11	100 nF minidisc	7	(YR75S)
C3	see additional parts below		
C4	1 µF 63 V min electrolytic	1	(YY31J)
C10,12	10 µF 16 V min electrolytic	2	(YY34M)
C13–18	2n2F ceramic	6	(WX72P)
C19–24	22 µF 16 V min electrolytic	6	(YY36P)

Semiconductors

IC1	TLC555C	1	(RA76H)
IC2,3	LF347N	2	(WQ29G)
IC4	MF10CN	1	(QY35Q)

Miscellaneous

	DIL socket 8-pin	1	(BL17T)
	DIL socket 14-pin	2	(BL18U)
	DIL socket 20-pin	1	(HQ77J)
P1–26	1 mm PCB pins	1	(FL24B)
	constructors' guide	1	(XH79L)

Additional parts

R1	1 k	1	(M1K)
R1	10 k	1	(M10K)
R2	4k7	1	(M4K7)
R2	47 k	1	(M47K)
RV1	100 k hor encl preset	1	(UH06G)
RV1	1 M hor encl preset	1	(UH09K)
C3	100 pF 1% polystyrene	1	(BX46A)
C3	1000 pF 1% polystyrene	1	(BX56L)

SSM2044 voltage controlled filter

The SSM2044 is a low cost, 4-pole voltage controlled filter integrated circuit which is ideal for use as an electronic low pass filter. It is also possible to use the device as a voltage controlled sinewave oscillator. Figure 2.31 shows the integrated circuit pinout and Table 2.10 shows the typical electrical characteristics of the device.

Integrated circuit description

The SSM2044 uses a unique filtering technique to provide low noise operation and a high rejection of control signals with an extended control range. Figure 2.32 shows typical filter responses obtainable using the integrated circuit. The differential signal inputs will accept signals up to ±18 V peak to peak.

Parameter	Conditions	Min	Typ	Max
Positive supply voltage		+5 V	+15 V	+18 V
Negative supply range		−5 V	−15 V	−18 V
Positive supply current	Pin 13 at 0 V	1.0 mA	1.4 mA	2.0 mA
Negative supply current	Pin 13 at 0 V	4.5 mA	6.2 mA	8.0 mA
Q control threshold voltage		400 mV	500 mV	
Frequency control range		10000/1	50000/1	
Frequency control input range		120 mV	+1 mV	80 mV

Table 2.10   IC electrical characteristics

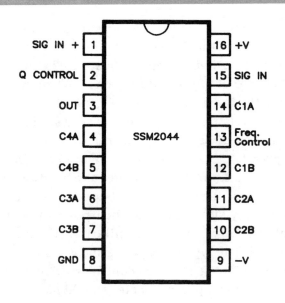

Figure 2.31   IC pinout diagram

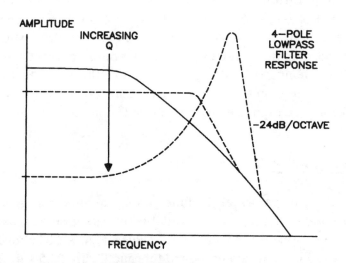

Figure 2.32   Typical SSM2044 filter responses

Audio IC projects

The effective Q of the filter is determined by the current flow into pin 2 of the device. When the Q control current reaches a critical value (approximately 425 µA) oscillation will occur at the cut-off frequency. When used as an oscillator the integrated circuit is capable of producing a comparatively pure sinewave. For all Q settings below the oscillation threshold the final roll-off at high frequencies approximates –24 dB/octave. Figure 2.33 shows the resonance of a 4-pole low pass filter as a function of feedback or Q control current. It can be seen that the rate of change is very slow when the Q current is low, but increases rapidly as the oscillation threshold is approached. The type of response shown can be a problem when designing a Q panel control with the right *feel*. Ideally the control potentiometer should have a characteristic which is a reciprocal of this response. One way to approximate this response is to connect a logarithmic potentiometer in reverse of the standard configuration. To obtain better resolution a resistor equal to one third of the value of the pot can be connected in series with the 0 V end to ground, thus discarding the lower 25% of the Q response curve, where little change is evident. The sense of the Q control is such that minimum resonance is achieved at 0 V and the resonance increases with positive Q control current.

Kit available

A kit of parts, which includes a high quality fibreglass printed circuit board, is available for a general purpose application circuit using the SSM2044. Figure 2.34 shows the circuit diagram of the module and Figure 2.35 shows the legend.

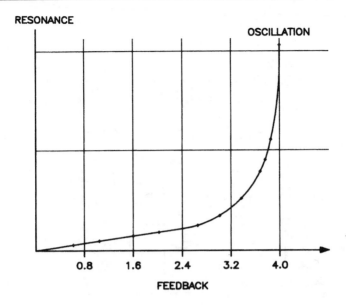

Figure 2.33   Resonance verses Q control current

Additional components have been included in the design to allow the circuit to operate from a single rail supply. The module requires a single power supply of between 12 V and 30 V which is capable of supplying up to 50 mA. As always, it is important that the power supply is adequately smoothed to prevent the introduction of mains hum onto the power supply rails. When using a single supply, power supply connections are made to P1(+V) and P3(0 V). Figure 2.36 shows the wiring information for single supply operation.

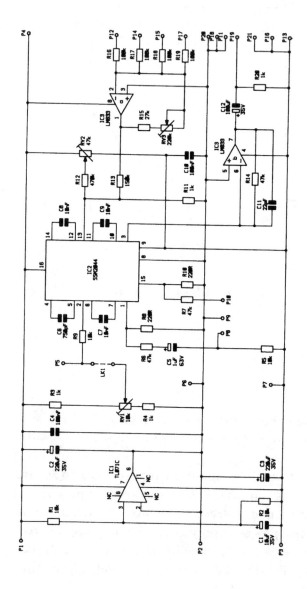

Figure 2.34 Circuit diagram

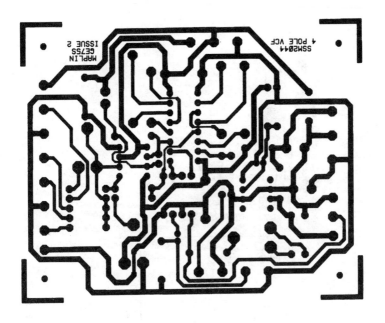

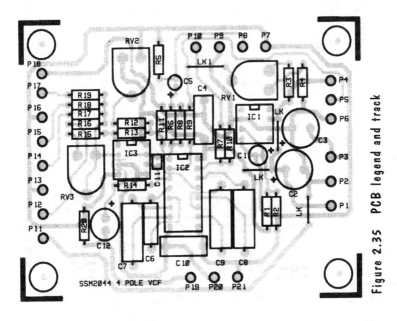

Figure 2.35 PCB legend and track

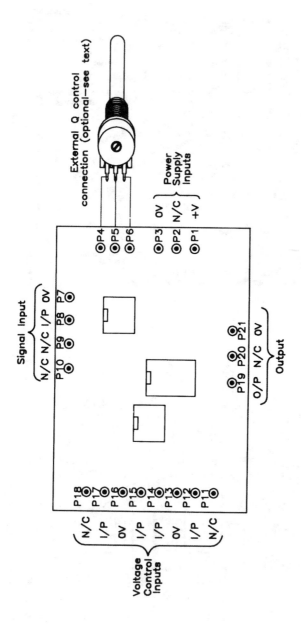

Figure 2.36 Wiring diagram for single supply operation

It is equally possible to power the circuit from a split rail power supply of between ±6 V and ±15 V, and when the module is used in this way some components (resistors R1, R2, R5, and R20, capacitor C1, and integrated circuit IC1) should be omitted, and wire links fitted in place of capacitors C5 and C12. When using a split supply, power supply connections are made to P1(+V), P2(0 V) and P3(V). Wiring information for this type of configuration is shown in Figure 2.37.

The circuit can use either single ended or balanced inputs and provides a single ended output. The non-inverting input and the output have facilities for capacitive coupling. If a single ended input is required, then input connections should be made to P8 with the 0 V return to P7 or P9 as appropriate. For a balanced input, connections should be made to P8(+i/p), P9(0 V) and P10(–i/p). Note: the balanced input configuration is practical only when using a split supply.

A total of four voltage control input pins are provided for frequency control (P12, P14, P15 and P17), all of which are connected via 100 k resistors to the inverting input of IC3(a), which effectively acts as a summing amplifier. Preset resistor RV2 is used to set the centre point of the frequency range, while preset RV3 adjusts the sensitivity of the voltage control inputs.

Wire link LK1 allows the choice of either an on-board Q control preset (RV1) or an external potentiometer. If LK1 is fitted, Q control is via the on-board preset; however, an external Q control potentiometer may be connected to P4, P5 and P6 and in this case LK1, preset RV1, and resistors R3 and R4 must be omitted. A suitable value for the potentiometer is 10 kΩ.

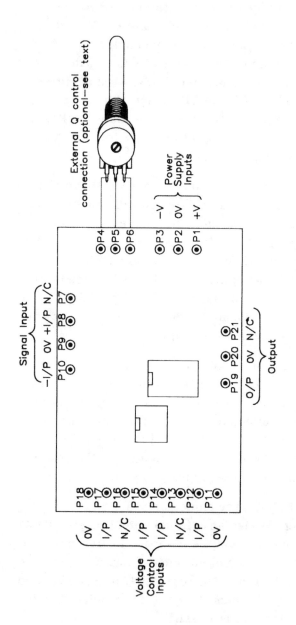

Figure 2.37 Wiring diagram for split supply operation

As mentioned, some of the components are only required in certain configurations and to illustrate this, Table 2.11 shows the appropriate components together with the different circuit options available. The table also shows which of the options are available for use with split (balanced) or single supplies only. In addition to LK1 there are three other wire links on the printed circuit board (marked LK); these are all fitted independent of the options chosen. Please note: if all of the components shown

	Single supply	Split supply	On board Q control	External Q control	Single ended input	Balanced input
LK1			yes	no		
R1	yes	no				
R2	yes	no				
R3			yes	no		
R4			yes	no		
R5	yes	no				
R7					no	yes
R20	yes	no				
RV1			yes	no		
C1	yes	no				
C5	yes	linked				
C12	yes	linked				
IC1	yes	no				
Single ended input	yes	yes				
Balanced input	no	yes				

Table 2.11 Construction information for different options

in the parts list are fitted, the module is then configured to operate from a single rail power supply, with Q control via preset RV1 and with a single ended input.

Applications

The module may be used in many different applications requiring a voltage controlled low pass filter or oscillator. Operating frequency ranges can be changed by fitting different values in place of capacitor C6. Optimum performance will usually be achieved using the highest specified supply voltage as this allows improved dynamic range.

By applying the output of a low frequency ramp or triangular wave oscillator to the control voltage input, the module could be used to form part of a sweep oscillator circuit; this can be used in conjunction with an oscilloscope to display filter frequency responses. The oscilloscope may then be triggered using the same ramp/triangle wave form used to drive the voltage controlled oscillator. A typical set-up for this type of system is shown in Figure 2.38, Table 2.12 shows the specification of the prototype SSM2044 module.

Power supply voltage	Single supply	12 V–30 V
	Split supply	±6 V–±18 V
Power supply current(Quiescent)	15 mA at 30 V	
Operating frequency range	(Cut-off frequency)	10 Hz–50 kHz
Voltage control input range		0 V–30 V

Table 2.12   Specification of prototype. (Power supply = 30 V unless otherwise specified)

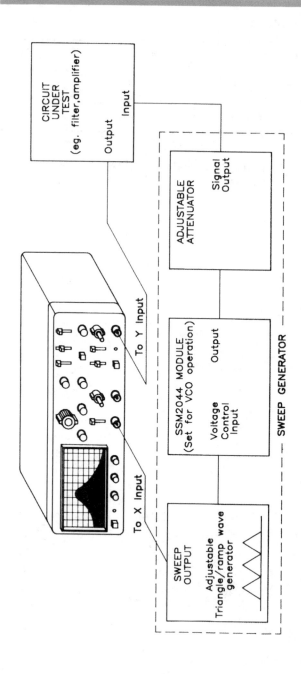

Figure 2.38 Basic block diagram of a sweep generator incorporating the SSM2044 module

SSM2044 4-pole voltage controlled filter parts list

Resistors — All 0.6 W 1% metal film

R1,2,5,9	10 k	4	(M10K)
R3,4, 11,20	1 k	4	(M1K)
R6,7,14	47 k	3	(M47K)
R8,10	220 Ω	2	(M220R)
R12	470 k	1	(M470K)
R13	150 k	1	(M150K)
R15	27 k	1	(M27K)
R16,17, 18,19	100 k	4	(M100K)
RV1	10 k hor encl preset	1	(UH03D)
RV2	47 k hor encl preset	1	(UH05F)
RV3	220 k hor encl preset	1	(UH07H)

Capacitors

C1	100 µF 35 V min electrolytic	1	(JL05F)
C2,3	220 µF 35 V PC electrolytic	2	(JL22Y)
C4,10	100 nF polyester	2	(BX76H)
C5	1 µF 63 V min electrolytic	1	(YY31J)
C6	750 pF polystyrene 1%	1	(BX55K)
C7,8,9	10,000 pF polystyrene 1%	3	(BX86T)
C11	22 pF ceramic	1	(WX48C)
C12	100 µF 35 V PC electrolytic	1	(JL19V)

Audio IC projects

Semiconductors

IC1	TL071CN	1	(RA67X)
IC2	SSM2044P	1	(UL19V)
IC3	LM833N	1	(UF49D)

Miscellaneous

P1–21	1 mm PCB pins	1	(FL24B)
	DIL socket 8-pin	2	(BL17T)
	DIL socket 16-pin	1	(BL19V)
	printed circuit board	1	(GE75S)
	constructors' guide	1	(XH79L)

3 Odds and ends

Featuring:

MN3004 bucket brigade delay line

The MN3004 is a 512 stage, low noise, bucket brigade delay line. Figure 3.1 shows the integrated circuit pinout and Table 3.1 shows some typical electrical characteristics. The device features low insertion loss, a signal-to-noise ratio of 85 dB, and no gate back-bias is required. Figure 3.2 shows the MN3004 block diagram, while Figure 3.3 shows its internal schematic.

Analogue signals, in the audio band, can be delayed by 2.56 ms to 25.6 ms by adjusting the clock frequency. The device is ideally suited for processing audio signals to produce an artificial delay in public address systems.

Parameter		Min	Typ	Max	Unit
Drain supply voltage	V_{DD}	−14	−15	−16	V
Gate supply voltage	V_{GG}		$V_{DD} + 1$		V
Clock frequency	f_{CP}	10		100	kHz
Signal delay time	t_D	2.56		25.6	ms
Signal frequency response	f_i			$0.3 \times f_{CP}$	kHz
Insertion loss	L_i	−4	1.5	4	dB
Total harmonic distortion	THD		0.4	2.5	%
Signal-to-noise ratio	S/N f_{CP} = 100 kHz				
	weighted by				
	A curve		85		dB

Table 3.1   MN3004 typical electrical characteristics

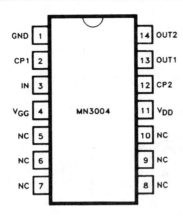

Figure 3.1   MN3004 pinout

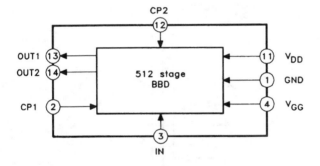

Figure 3.2   IC block diagram

Bucket brigade principle

Sampled values of the analogue signal to be delayed are stored in the form of charges on a series of capacitors. Between each capacitor is a switch that transfers the charge from one capacitor to the next, upon command

123

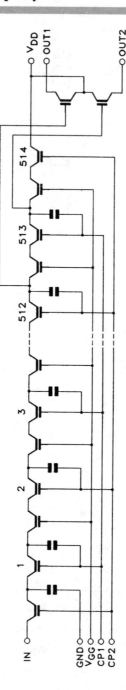

Figure 3.3 MN3004 internal schematic

of a clock pulse. Using the old analogy of the *fire-fighting* method, in which buckets of water were passed along a line, from man to man, a delay line of this sort is known as a bucket-brigade.

As each capacitor cannot take-up a new charge until it has passed on its previous one, only half the capacitors carry information at any one-time and the ones in between are empty. Starting from the condition shown in Figure 3.4(a), the transfer proceeds in two stages:

● in the first stage, bucket 1 empties into bucket 2, and bucket 3 into bucket 4,

● in the second stage, bucket 2 empties into bucket 3, and bucket 4 into bucket 5.

Two antiphase clock signals are therefore required: one to empty the even-numbered buckets, and the other to empty the odd-numbered buckets.

There is, however, a practical drawback to the above method, as the buckets in which the samples are stored must empty *completely* during each transfer. In practice, owing to internal resistance of the capacitors, complete discharge is difficult to ensure. So instead, the system illustrated in Figure 3.5 is employed.

Figure 5(a) indicates the initial condition; buckets 1 and 3 contain samples, and buckets 2 and 4 are full. During the first transfer, bucket 2 fills bucket 1, and bucket 4 fills bucket 3. What remains in bucket 2 is now equal to the original contents of bucket 1, and what remains in

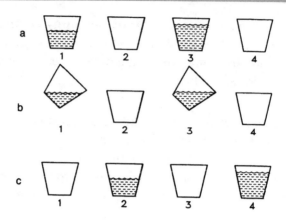

Figure 3.4 Simplified scheme of sample transfer: samples in buckets 1 and 3 (a) are poured into empty buckets 2 and 4 (b) causing the samples to move from left to right (c)

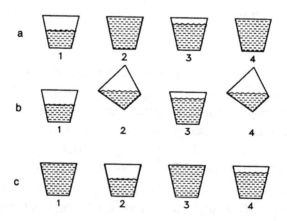

Figure 3.5 Practical scheme of sample transfer: buckets 1 and 3 contain samples (a) and, when full buckets 2 and 4 are poured into them (b), the samples move from left to right (c)

bucket 4 is equal to the original contents of bucket 3. During the next transfer, bucket 3 fills bucket 2, and bucket 5 fills bucket 4. So, as the buckets empty from right to left, the sampled quantities move from left to right.

Kit available

A kit of parts is available to build a versatile application circuit using the MN3004. The kit includes a high quality fibreglass printed circuit board with screen printed legend to aid construction. Figure 3.6 shows the circuit diagram while Figure 3.7 shows the printed circuit board legend and track. A block diagram of the module is shown in Figure 3.8.

Setting-up

The module requires a single +15 V supply that is capable of delivering at least 40 mA. It is important that the power supply is adequately smoothed and regulated to prevent any mains-derived noise from entering the system via the supply rails. Power supply connections are made to P1(+15 V) and P2(0 V). Signals to the delay line are applied via P3(input) and P4(0 V), and signals are output from the module via P5(output) and P6(0 V). Figure 3.9 shows the wiring diagram.

Audio IC projects

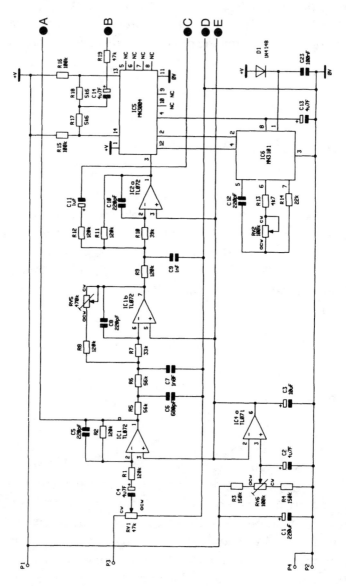

Figure 3.6(a)

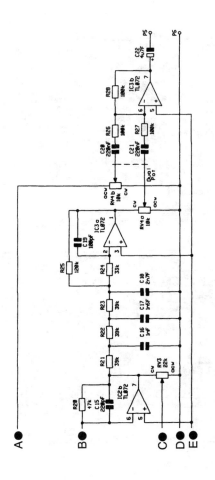

Figure 3.6(b)   Circuit diagram

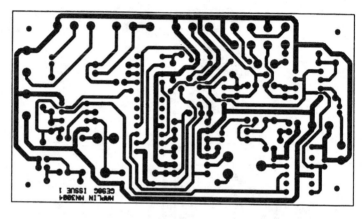

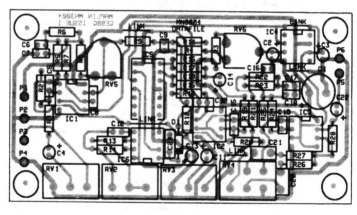

Figure 3.7 PCB legend and track

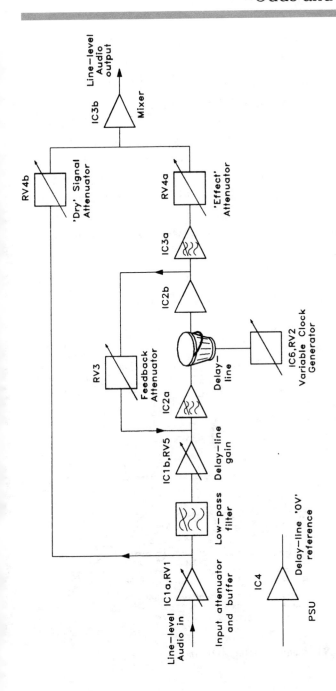

Figure 3.8 Module block diagram

Delay line biasing

The d.c. offset at the signal input of the MN3004 must be adjusted for minimum distortion. Starting from a level of half the supply voltage (preset RV6 set to mid-way), a potential change of up to 2 V may be necessary. Connect the module as per Figure 3.9, set preset RV5 to its mid-way position, potentiometer RV3 fully anti-clockwise (no feed-back) and potentiometer RV4 fully clockwise (pure effect output). Increase the input level by rotating potentiometer RV1 clockwise, followed by preset RV5,

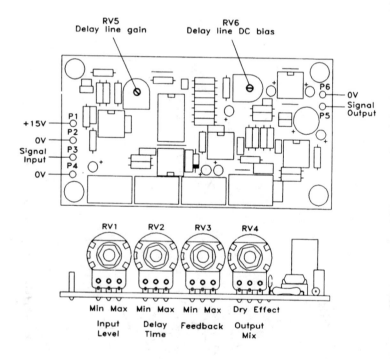

Figure 3.9   Module wiring diagram

until the output signal is distorted. Next, adjust preset RV6 for minimum distortion, and increase the input level for a second adjustment. Optimise the drive margin this way until no further improvement is noted.

Delay line loop gain

With potentiometer RV2 (delay) set to mid-range and potentiometer RV3 (feedback) fully anti-clockwise connect an input signal of about 1 V peak-to-peak and monitor the output signal. Adjust preset RV5 so that equal output levels are obtained with potentiometer RV4 (mix) fully clockwise and fully anti-clockwise. This can be done with test equipment or simply *by ear* using a tape or digital signal source.

Applications

The module may be used in many different applications requiring a signal delay time of up to 25.6 mS including musical instruments, communications systems and time compression.

As audio signals enter the module they split into two paths, which are re-united again near the output. One of these paths is a direct link from the input of the module to the mixer stage (potentiometer RV4 and integrated circuit IC3(b)); the other path is via the MN3004 delay line. Note that, as the MN3004 clock frequency must be at least 3 times the maximum input signal frequency, the delayed signal has a limited bandwidth of 3.1 kHz. In practice, this restricted bandwidth is hardly a problem since higher frequencies are also attenuated with natural echoes and reverberation. By varying the signal delay and the mixture of direct and delayed signals, a variety of interesting effects can be obtained. Here are a few ideas to try:

● with equal levels of direct and delayed signals, and a few milliseconds of delay, a *double-tracking* effect is produced. This makes a single input sound like a pair of independent, but time synchronised, outputs. Using this effect a single voice sound can be made to sound like a duet, and a duet made to sound like a quartet.

● with a reduced level of delayed signal in the mix, and a reasonable length delay time, a simple echo effect is obtained. The audio sounds as if it were being played in a softly furnished room with a single hard wall facing the sound source.

● when equal levels of direct and delayed signals are used with a long delay time and almost maximum feedback, the sounds seem as if they are being played in a hard-faced cave. The apparent dimensions of this chamber can be varied through the delay-time control, while the *hardness* of the chamber can be altered by the feedback control. The apparent sounds can be varied from those of a hard cave, to a small church, or even down to a large but well furnished lounge.

● when equal levels of mixing are used with a short delay time and a large amount of feed-back, all sounds give the impression that they are being played inside a small-diameter, hard-faced pipe. The dimensions of the pipe can be varied with the delay time control, while the hardness of the pipe is variable through the feedback control. This allows the sounds to be varied from those of a sewer pipe to a bucket.

Finally, Table 3.2 shows the specification of the prototype MN3004 module.

Parameter	Min	- Typ	Max	Unit
Supply voltage	14	15	16	V
Supply current		36		mA
Input signal level (RV1 and RV4 fully clockwise)		1	3	V r.m.s.
Input signal level (RV1 fully clockwise, RV4 fully anti-clockwise)		1	8	V r.m.s.
Input impedance		47		kΩ
Delay time	2		24	ms
Dry signal path frequency response	4		15 k	Hz
Delay signal path frequency response	4		3.1 k	Hz
Output signal level		1		V r.m.s.
Output impedance		600		Ω

Table 3.2 Specification of prototype (V_{cc} = 15 V)

MN3004 parts list

Resistors — All 0.6 W 1% metal film (unless specified)

R1,2,8,9, 11,12,25	120 k	7	(M120K)
R3,4	150 k	2	(M150K)
R5,6	56 k	2	(M56K)
R7,24	33 k	2	(M33K)
R10,21, 22,23	39 k	4	(M39K)
R13	4k7	1	(M4K7)
R14	22 k	1	(M22K)
R15,16, 26,27,28	100 k	5	(M100K)
R17,18	5k6	2	(M5K6)
R19,20	47 k	2	(M47K)
RV1	47 k min pot log	1	(JM78K)
RV2	100 k min pot lin	1	(JM74R)
RV3	22 k min pot lin	1	(JM72P)
RV4	10 k min dual pot lin	1	(JM81C)
RV5	470 k hor encl preset	1	(UH08J)
RV6	100 k hor encl preset	1	(UH06G)

Capacitors

C1	220 µF 35 V PC electrolytic	1	(JL22Y)
C2,4,13, 14,22	4µ7F 35 V min electrolytic	5	(YY33L)
C3	10 µF 16 V min electrolytic	1	(YY34M)
C5,8,10, 12,15	220 pF ceramic	5	(WX60Q)

C6	680 pF ceramic	1	(WX66W)
C7	1n8F ceramic	1	(WX71N)
C9,16	1 nF ceramic	2	(WX68Y)
C11	1 µF 63 V min electrolytic	1	(YY31J)
C17	1n5F ceramic	1	(WX70M)
C18	2n7F ceramic	1	(WX73Q)
C19	100 pF ceramic	1	(WX56L)
C20,21	220 nF monores cap	2	(RA50E)
C23	100 nF 16 V minidisc	1	(YR75S)

Semiconductors

D1	1N4148	1	(QL80B)
IC1,2,3	TL072CN	3	(RA68Y)
IC4	TL071CN	1	(RA67X)
IC5	MN3004	1	(UM64U)
IC6	MN3101	1	(UM66W)

Miscellaneous

P1–6	1 mm PCB pins	1	(FL24B)
	8-pin DIL socket	5	(BL17T)
	14-pin DIL socket	1	(BL18U)
	PC board	1	(GE98G)
	constructors' guide	1	(XH79L)

MN3011 bucket brigade delay line

The MN3011 is a 3328 stage bucket-brigade delay line with 6 tap outputs. Analogue signals, in the audio band, can be delayed by 1.98 ms to 166.4 ms by adjusting the clock frequency and making connection to the relevant output (for details see Table 3.3). Natural reverberation effects can also be produced by summing two or more of the six outputs. Figure 3.10 shows the integrated circuit's pinout and Table 3.4 shows some typical electrical characteristics for the device.

Reverberation

Most electronic musical instruments (and acoustic instruments when played in a studio environment) are not based on sound created in a resonant room, and thus

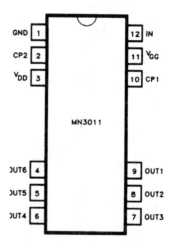

Figure 3.10   IC pinout

Terminal No.	Symbol	Description
1	GND	
2	CP2	Clock pulse 2
3	V_{DD}	Supply voltage of −15 V
4	OUT 6	Output of stage 3328 and 3329
5	OUT 5	Output of stage 2790 and 2791
6	OUT 4	Output of stage 1726 and 1727
7	OUT 3	Output of stage 1194 and 1195
8	OUT 2	Output of stage 662 and 663
9	OUT 1	Output of stage 396 and 397
10	CP1	Clock pulse 1
11	V_{GG}	Gate bias terminal
12	IN	Signal input terminal

Table 3.3 Terminal description

Parameter		Min	Typ	Max	Unit
Supply voltage	V_{DD}	−14	−15	−16	V
Gate supply voltage	V_{GG}		V_{DD} + I		V
Clock frequency	f_{CP}	10		100	kHz
Signal delay time					
OUT I Terminal	t_{DI}	I.98		19.8	ms
OUT 2Terminal	t_{D2}	3.31		33.1	ms
OUT 3 Terminal	t_{D3}	5.97		59.7	ms
OUT 4 Terminal	t_{D4}	8.63		86.3	ms
OUT 8 Terminal	t_{D5}	13.95		139.5	ms
OUT 6 Terminal	t_{D6}	16.64		166.4	ms
Signal frequency response	f_i			$0.3 \times f_{CP}$	kHz
Insertion loss	L_i	−4	0	4	dB
Total harmonic distortion	THD		0.4	2.5	%
Signal-to-noise ratio	S/N f_{CP} = I00 kHz weighted by A curve		76		dB

Table 3.4 Typical electrical characteristics

Audio IC projects

produce a *flat* sound. Reverberation may be added by electronic means to add warmth to the sound of these instruments. A single pulse of sound (e.g. a handclap) will send a set of waves out across a room and, when they strike objects within the room, the direction and intensity of the sound waves will be changed. These changes in direction are caused by the sound waves being reflected, and it is these reflections that are the main cause of reverberation as the sound dies away. The attenuation and decay-time of the reverberation is determined by the air and absorbent surfaces, as shown in Figure 3.11.

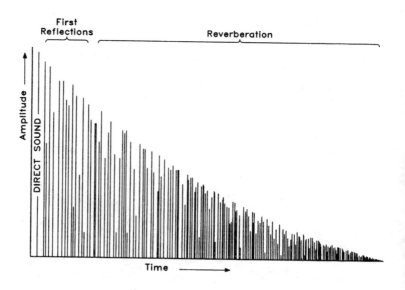

Figure 3.11 The profile of reflections of a single pulse of sound

All electronic systems used for adding reverberation depend on the same basic principle; the signal is delayed for a short time and part of it is fed back to the input. By carefully controlling the delay and amount of feedback, an effective *reverb* can be produced. In reality, all that is being done is to recirculate a series of short echoes. If this is taken to extreme levels, oscillation results; this is hardly surprising when you consider that signals are being added, one on top of the other. However, when proper attention is paid to setting the levels, the sound can be quite realistic.

Figure 3.12 shows a simple reverberation unit based on *bucket brigade* memory. The memory is essentially a series of sample-and-hold circuits, each of which consists of an electronic switch and capacitor. The analogue signals stored in the capacitors are sampled under the control of a central clock signal. At each clock pulse the samples are shifted one capacitor to the right, hence the name *bucket brigade* (the predecessor of today's fire brigade). However, a lifelike reverberation effect can only be achieved by using multiple delays of non-related durations.

Kit available

A kit of parts including a high quality fibreglass printed circuit board with screen printed legend is available to enable construction of a general purpose module based on the MN3011. Figure 3.13 shows the circuit diagram of the module and Figure 3.14 shows the block diagram; the printed circuit board layout is shown in Figure 3.15.

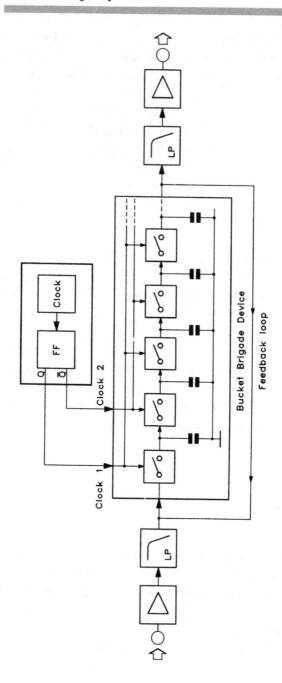

Figure 3.12   Block diagram of a simple BBD reverberation unit

Odds and ends

The MN3011 has six different and non-related delays, while the attenuation of each of these is adjustable to give an optimum room simulation (presets RV4 to RV9). Recirculation of the echoes is produced by feeding back part of the longest echo to the input (preset RV3).

The module requires a single +15 V supply that is capable of delivering at least 40 mA. It is important that the power supply is adequately smoothed and regulated to prevent any mains-derived noise from entering the system via the supply rails. Power supply connections are made to P1 (+15 V) and P2 (0 V). Signals to the reverb unit are applied via P3 (input) and P4 (0 V), and signals are output from the module via P5 (output) and P6 (0 V). Figure 3.16 shows the wiring diagram. Note that the cases of potentiometer RV1, 2 and 11 are connected to 0 V through a wire link.

Delay line biasing

The d.c. offset at the signal input of the MN3011 must be adjusted for minimum distortion. Starting from a level of half the supply voltage (preset RV12 set to mid-way), a potential offset of up to 2 V may be necessary. Connect the module as per Figure 3.16, set preset RV12 to its mid-way position, potentiometer RV2 and preset RV4 fully clockwise, and all other presets fully anti-clockwise. Increase the input level by rotating potentiometer RV1 clockwise, followed by preset RV10, until the output signal is distorted. Next, adjust preset RV12 for minimum distortion, and increase the input level for a second adjustment. Optimise the drive margin this way until no further improvement is noted.

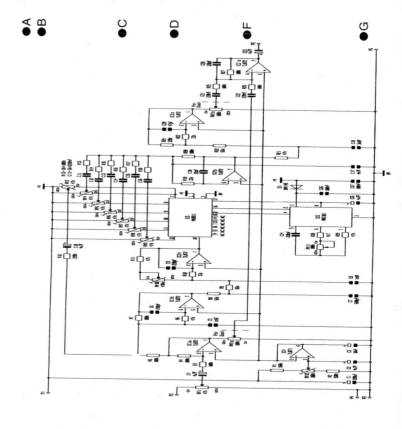

Figure 3.13(a)

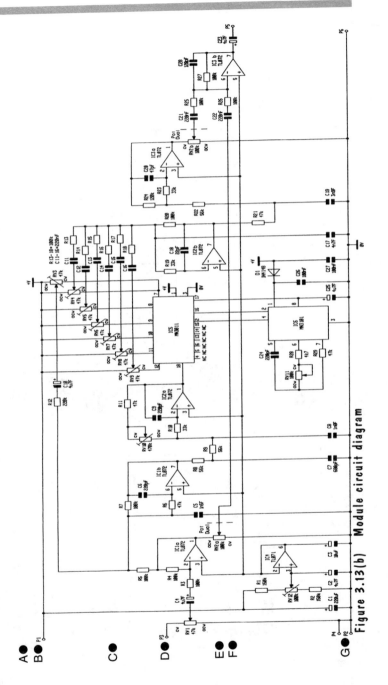

Figure 3.13(b)   Module circuit diagram

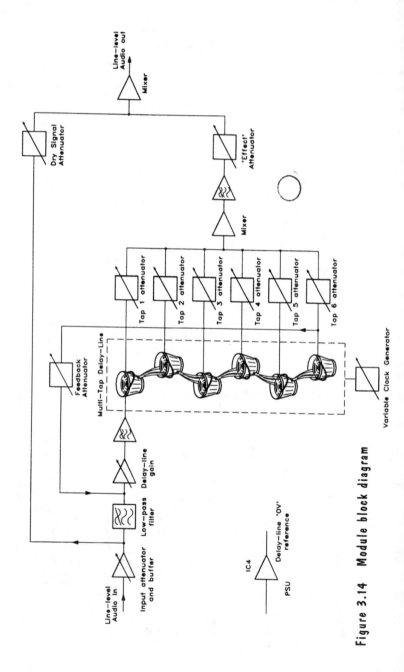

Figure 3.14 Module block diagram

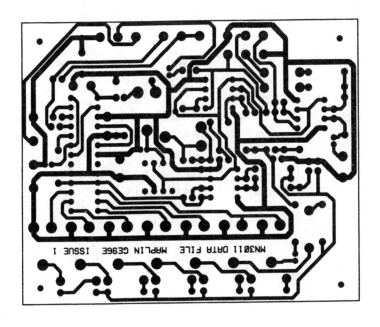

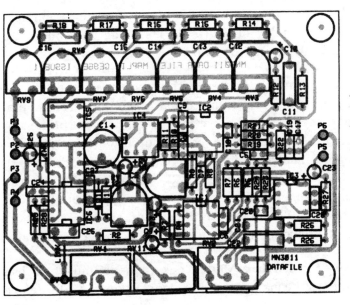

Figure 3.15   PCB legend and track

Delay line loop gain

With potentiometer RV3 (feedback) fully anti-clockwise, potentiometer RV11 (reverb time) and presets RV4 to RV9 set to their mid-way position, connect an input signal of about 1 V peak-to-peak and monitor the output signal. Adjust preset RV10 so that equal output levels are obtained with potentiometer RV2 (mix) fully clockwise and fully anti-clockwise. This can be done with test equipment or simply *by ear* using a tape or digital signal source.

Applications

The module may be used in many different applications requiring a natural reverb effect, or even a single or multiple delay of up to 166 ms using only one of the tapped outputs from the MN3011.

Audio signals, on entering the module, are split into two paths, which are re-united near the output. The simplest of these paths is a direct link from the input of the module to the output mixer stage (potentiometer RV2(a) and operational amplifier IC3(b)); the second path is via the MN3011 delay line. Presets RV4 to RV9 set how much signal is mixed from each of the six tapped outputs from the MN3011 and sent to the output, and preset RV3 controls the amount of echo sent back for recirculation.

Note that, as the MN3011 clock frequency must be at least 3 times the maximum input signal frequency, the delayed

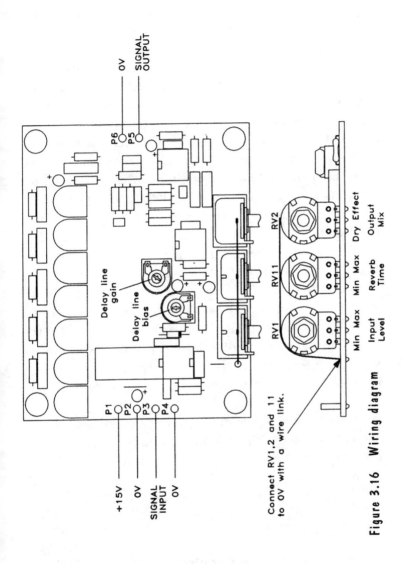

Figure 3.16 Wiring diagram

signal has a limited bandwidth of 3.3 kHz. This bandwidth restriction is of little consequence to the audible results as higher frequencies in a *live* environment are quickly absorbed and do not, therefore, reverberate.

Applications for this module include adding reverb to a domestic hi-fi environment, thus recreating the spaciousness or reverberation experienced in a concert hall or theatre. For example, the system could be entirely separate with its own amplifier and loudspeakers. The level of the reverberated sound should be about half of that from the main system, and the loudspeakers should be positioned such that they cannot be heard as an identifiable sound source. Alternatively, the reverberated signals can be mixed with the dry audio source and fed to the main amplifier. In this case, separate loudspeakers and amplifier are not necessary.

By varying the delayed/reverb signal mix (presets RV4–RV9), feedback (preset RV3), reverb time (potentiometer RV11) and mixture of direct and delayed/reverberated signals (potentiometer RV2), a variety of interesting effects can be obtained. Here are a few ideas to try:

● with equal levels of only one delay tap and direct signals, coupled with a few milliseconds of delay, a *double-tracking* effect is produced. This makes a single input sound like a pair of independent, but time-synchronised, outputs. By adding the delayed signals of further tap outputs the duet can be turned into a quartet, or even an octet, etc.

● when equal levels of direct and a mix of delayed signals are used with a long reverb time, and almost maximum feedback, the sounds seem as if they are being played in a large cathedral. The apparent dimensions of this *chamber* can be varied through the reverb-time control, while the *hardness* of the chamber can be altered by the feedback control. The sounds can be varied from those of a large cavern, down to a well furnished lounge.

● when equal levels of mixing are used with a short reverb time and a small quantity of feedback, the impression is given that all sounds are being played inside a small, hard-faced pipe. The size of the *pipe* can be altered with the reverb-time control and the hardness of the pipe is variable through the feedback control, allowing sounds to be varied from those of a sewer pipe to a bucket.

Finally, Table 3.5 shows the specification of the prototype MN3011 module.

Parameter	Min	Typ	Max	Unit
Supply voltage	14	15	16	V
Supply current		40		mA
Input signal level (RV1 and RV2 fully clockwise)		1	2.5	V r.m.s.
Input signal level (RV1 fully clockwise, RV2 fully anti-clockwise)		1	8	V r.m.s.
Input impedance		47		kΩ
Delay time	1.7		160	ms
Dry signal path frequency response	7		23 k	Hz
Delay signal path frequency response	8		3.3 k	Hz
Output signal level		1		V r.m.s.
Output impedance		600		Ω

Table 3.5 Specification of prototype

MN3011 bucket-brigade delay-line parts list

Resistors — All 0.6 W 1% metal film (unless specified)

R1,2,3, 4,5,7, R13–18,	150 k	2	(M150K)
20,25–27	100 k	14	(M100K)
R6,11, 21,29	47 k	4	(M47K)
R8,9,22	56 k	3	(M56K)
R10,19, 23	33 k	3	(M33K)
R12	220 k	1	(M220K)
R24	120 k	1	(M120K)
R28	4k7	1	(M4K7)
RV1	47 k min pot log	1	(JM78K)
RV2	100 k min dual pot lin	1	(JM82D)
RV3–9	47 k hor encl preset	7	(UH05F)
RV10	470 k hor encl preset	1	(UH08J)
RV11	100 k min pot lin	1	(JM74R)
RV12	100 k hor encl preset	1	(UH06G)

Capacitors

C1	220 µF 35 V PC electrolytic	1	(JL22Y)
C2,4,10, 23,25	4µ7F 35 V min electrolytic	5	(YY33L)
C3	10 µF 16 V min electrolytic	1	(YY34M)
C5	1n5F ceramic	1	(WX70M)
C6,9,24	220 pF ceramic	3	(WX60Q)
C7	680 pF ceramic	1	(WX66W)

C8	1n8F ceramic	1	(WX71N)
C11–16, 21,22	220 nF polylayer	8	(WW45Y)
C17	4n7F ceramic	1	(WX76H)
C18	22 pF ceramic	1	(WX48C)
C19	3n9F ceramic	1	(WX75S)
C20	47 pF ceramic	1	(WX52G)
C26,27	100 nF 16 V minidisc	2	(YR75S)
C28	120 pF ceramic	1	(WX57M)

Semiconductors

D1	1N4148	1	(QL80B)
IC1,2,3	TL072CN	3	(RA68Y)
IC4	TL071CN	1	(RA67X)
IC5	MN3011	1	(UM65V)
IC6	MN3101	1	(UM66W)

Miscellaneous

P1–6	1 mm PCB pins	1	(FL24B)
	8-pin DIL socket	5	(BL17T)
	18-pin DIL socket	1	(HQ76H)
	PCB	1	(GE96E)
	constructors' guide	1	(XH79L)

153

MSM6322 speech pitch controller

The MSM6322 is a real-time audio pitch controller specifically designed for speech. Available solely in a surface mount package (with a size of only 16 x 12 mm!), the integrated circuit contains a fourth order low pass input filter and 8-bit analogue-to-digital converter, a 9-bit digital-to-analogue converter with a third order low pass output filter, as well as a microphone pre-amplifier. Over the time that this integrated circuit has been available to trade and industry it has found its way into many children's toys, transforming their voice into that of *Darth Vader* or *Mickey Mouse*! Figure 3.17 shows the integrated circuit pinout while Table 3.6 gives typical electrical characteristics for the device.

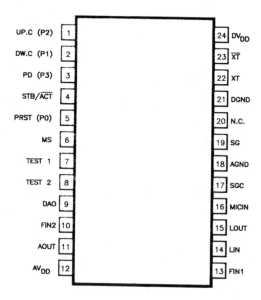

Figure 3.17   MSM6322 pinout

Parameter	Symbol	Condition	Min	Typ	Max	
Digital supply voltage	DV_{DD}	For D_{GND}	4.0	5.0	6.0	V
Analogue supply voltage	AV_{DD}	For A_{GND}	4.0	5.0	6.0	V
Operating current consumption	I_{DD}	In case of 4 MHz oscillator			10	mA
Standby current	I_{DS}	In case of 4 MHz oscillator with STB/ $\overline{ACT}$ = H			7	mA
Power down current	I_{DP}	In case of PD = H			1	mA
A-to-D conversion precision		$AV_{DD} = V_{DD} = 5\,V$			40	mV
D-to-A conversion precision		$AV_{DD} = V_{DD} = 5\,V$, No load			40	mV
Input impedance	MICIN	RI_{MICIN}		100		$M\Omega$
	LIN	RI_{LIN}		100		$M\Omega$
	FIN2	RI_{FIN2}		30		$M\Omega$
Output impedance	LOUT	RO_{LOUT}		15		$k\Omega$
	DAO	RO_{DAO}		10		$k\Omega$
	AOUT	RO_{AOUT}		15		$k\Omega$
	FIN1	RO_{FIN1}		15		$k\Omega$
Operating frequency	t_{CMAX}			4.0	4.5	MHz
Time between UP.C and DW.C pulses	t_{RUD}		30.72			ms
Pulse width of PRST, UP.C, DW.C pulses	t_{UDPW}		30.72			ms

Table 3.6 Electrical characteristics

Integrated circuit description

Figure 3.18, shows the block diagram of the MSM6322. Analogue and digital power supplies to the chip are completely isolated internally to reduce the chance of digital noise introducing itself into the analogue signal path.

Two operating modes are available:

● with the mode select (MS) pin set to logic low (0 V) the MSM6322 is placed in *up/down* mode. Pulses to the *Up Conversion* (UP.C) and *Down Conversion* (DW.C) pins raise and lower the pitch of the signal by one step per pulse,

● with the MS pin set high (5 V), the device is placed into *bin* mode. P0–P3 become binary inputs selecting the stage of pitch conversion.

Table 3.7 gives the pin functions common to both modes. Table 3.8 shows the relevant pin functions for *up/down* mode, and Table 3.9 for *bin* mode. Table 3.10 shows the relationship between scale stage, D-to-A sampling cycle, and low pass filter cut-off frequency.

Kit available

A kit of parts including a high quality fibreglass printed circuit board with printed legend is available as an aid to constructors, facilitating developments around this, sometimes tricky to mount, surface mount device. Figure 3.19 shows the circuit diagram of the module while Figure 3.20 shows the printed circuit board layout.

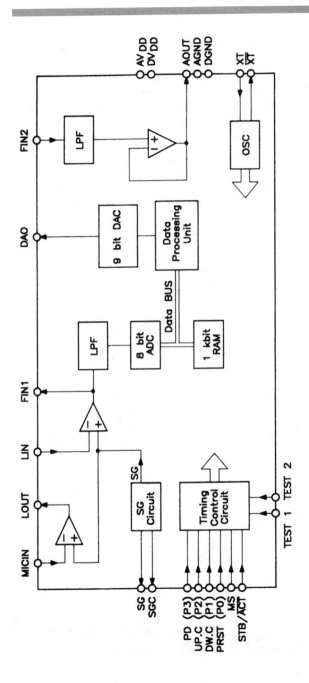

Figure 3.18   IC block diagram

Audio IC projects

Pin Name	Pin No	DI/AI/AO[1]	Function
MICIN	16	AI	Input to the microphone preamplifier. *Must* be capacitively coupled.
LOUT	15	AO	Output of the microphone preamplifier.
LIN	14	AI	Input to the line amplifier. *Must* be capacitively coupled to either LOUT or the line out signal from other audio sources.
FIN1	13	AO	Sets the input audio signal amplitude in combination with the LIN pin.
STB/$\overline{ACT}$	4	DI	Chip select pin. The processing is interrupted by stopping clocks other than the oscillator when the chip select pin is at the H level. The DAO pin outputs $\frac{1}{2}$ VDD for about 15 ms (in the case of 4 MHz oscillator) after the chip select pin is set to the L level.
TEST1	7	DI	Manufacturer's test pins.
TEST2	8		*Must* be connected to 0 V.
XT, $\overline{XT}$	22,23		Crystal oscillator connection pins.
SG,SGC	19,17		External capacitors are connected here to stabilise the internal analogue voltage references of $\frac{1}{2}$ AVDD.
DAO	9	AO	Output from the digital-to-analogue converter.
FIN2	10	AI	Input pin for internal low-pass filter (for output).
AOUT	11	AO	Output of low-pass filter (for output).
DGND	21		Digital power supply pins.
DVDD	24		
A_{GND}	18		Analogue power supply pins.
AVDD	12		

Note: DI (Digital input), AI (Analogue input), AO (Analogue output).

Table 3.7 Pin functions common to both modes

Pin name	Pin No.	Function
MS	6	Mode select pin, always connected to D_{GND}.
UP.C	1	Pulse input to raise pitch by one stage at a time.
DW.C	2	Pulse input to lower pitch by one stage at a time. Cyclic up or down operation is also possible. See Table 3.12 for scale stages.
PD	3	Power down pin. All clocks including the oscillator are stopped when the power down pin is set to the H level.
PRST	5	Pulse input sets the scale to stage 8 (no pitch change).

Table 3.8 Pin functions for up/down mode — all the above pins are digital inputs

The application circuit has an on-board 4 MHz clock, supply decoupling for both analogue and digital supplies. The gain of the microphone and line-level pre-amplifiers are set by two resistors in each case (the line-level pre-amplifier's gain is however, variable through preset RV1).

In constructing the module, take *great* care when soldering the MSM6322, which mounts on the *solder* side of the printed circuit board. It is recommended that the smallest possible size soldering iron bit is used (less than 1 mm), with 22 SWG solder. Using a pair of tweezers, place the integrated circuit squarely on all 24 pads. Solder the four corner pins, ensuring the device stays straight on the pads. Now solder the remaining pins, allowing several seconds between each solder joint for the integrated circuit to cool down.

Audio IC projects

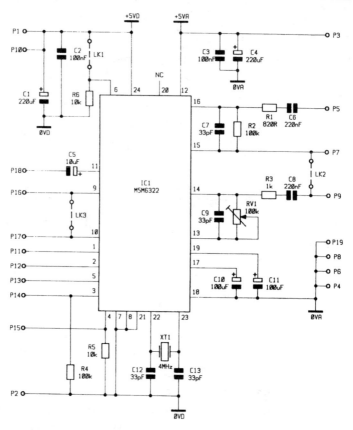

Figure 3.19   Circuit diagram

Pin name	Pin No.	Function
MS	6	Mode select pin, always connected to DV_{DD}.
P0	5	16 pitch stages are set by 4 bits of
P1	2	P3 (MSB) to P0 (LSB). Stages 0 (P3 = P2 = P1 = 0)
P2	1	to 16 (P3 = P2 = P1 = 1), as shown on Table 3.12,
P3	3	can be set.

Table 3.9   Pin functions for BIN mode — all the above pins are digital inputs.

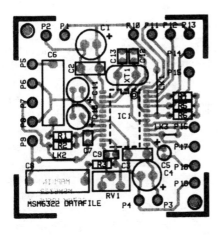

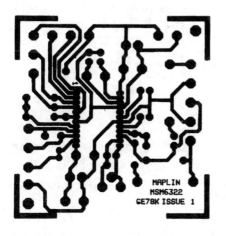

Figure 3.20 PCB legend and track

Audio IC projects

BIN mode settings				Scale stage	DA sampling cycle (µS) frequency (kHz)	LPF cut-off frequency (kHz)	Pitch change
P3	P2	P1	P0				
Not available				16	60/16.6	7.60	One octave up
1	1	1	1	15	71/14.0	7.60	Nine semi-tones up
1	1	1	0	14	76/13.1	5.70	Eight semi-tones up
1	1	0	1	13	80/12.5	5.70	Seven semi-tones up
1	1	0	0	12	90/11.1	5.70	Five semi-tones up
1	0	1	1	11	95/10.5	5.70	Four semi-tones up
1	0	1	0	10	101/9.90	4.56	Three semi-tones up
1	0	0	1	9	113/8.84	4.56	One semi-tone up
1	0	0	0	8	120/8.33	3.80	No pitch change
0	1	1	1	7	127/7.87	3.80	One semi-tone down
0	1	1	0	6	143/6.99	3.26	Three semi-tones down
0	1	0	1	5	151/6.62	3.26	Four semi-tones down
0	1	0	0	4	160/6.25	3.26	Five semi-tones down
0	0	1	1	3	180/5.55	2.85	Seven semi-tones down
0	0	1	0	2	190/5.26	2.53	Eight semi-tones down
0	0	0	1	1	202/4.95	2.53	Nine semi-tones down
0	0	0	0	0	227/4.40	2.07	One octave down

Table 3.10 Pitch conversion table

To use the device in *bin* mode it is necessary to insert link LK1, otherwise *up/down* mode is selected by resistor R6 pulling the mode select pin 6 low.

Connection information is given in Figure 3.21. Analogue and digital 5 V power supplies should be kept completely separate, meeting only at the power source to reduce the chance of digital noise introducing itself into the analogue signal path.

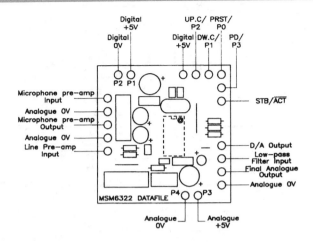

Figure 3.21 Wiring diagram

With reference to Figure 3.22, the gain of the microphone pre-amplifier is set to 122 (42 dB). The gain of the line-level pre-amplifier is variable, through preset RV1, from 0 to 100 (40 dB). If using the microphone pre-amplifier, link LK2 has to be installed to route the output from this pre-amplifier into the line-level pre-amplifier.

The output of the integrated circuit's digital-to-analogue converter is brought out to P16. This is *before* any output filtering has taken place. Note that this output has a +2.5 V d.c. offset, and will therefore need to be capacitively coupled to any external equipment. Link LK3, when installed, routes the output of the digital-to-analogue converter through the integrated circuit's low pass filter, the output of which is brought to P18.

163

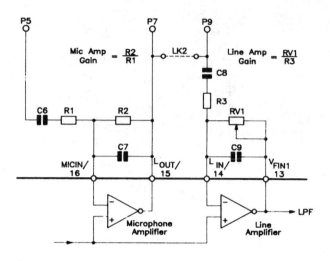

Figure 3.22   Setting the gain of the input amplifiers

The functions of P11 to P14 vary depending which mode the MSM6322 is set to, as shown in Table 3.11.

Pin Number	Up/down mode	BIN mode
P11	UP.C	P2
P12	DW.C	P1
P13	PRST	P0 (LSB)
P14	PD	P3 (MSB)

Table 3.11   Pin functions dependent on mode

In use

Interesting effects can be found by slowly increasing the line amplifier's gain via preset RV1, until distortion *just* starts to set in. This, coupled with a low pitch shift can generate very convincing *Dalek* effects. Also, in *up/down* mode, continuously pulsing the UP.C or DOWN.C inputs produces a rather strange effect.

As the MSM6322 does not boast a very high signal-to-noise ratio it may be beneficial to route the audio signal through a compander. A suitable circuit is shown in Figure 3.23.

The process of compansion can be broken down into two stages, compression (Figure 3.23(a)) and expansion (Figure 3.23(b)).

Compression involves reducing the dynamic range of the material that is being processed, so that, with a 2:1 compression ratio, if the input to the compressor increases by 12 dB, then the output of the compressor will increase by only 6 dB.

Conversely, expansion involves increasing the dynamic range, so that if the input to the expander increases by 6 dB, the output will increase by 12 dB, i.e. a 1:2 expansion ratio.

At the same time, noise introduced in the system will be rendered nearly inaudible on expansion as this unwanted signal is not subject to the initial compression treatment and is therefore expanded downwards below the lowest dynamics of the wanted audio signal. Figure 3.23(c) shows how to connect to the printed circuit board.

Audio IC projects

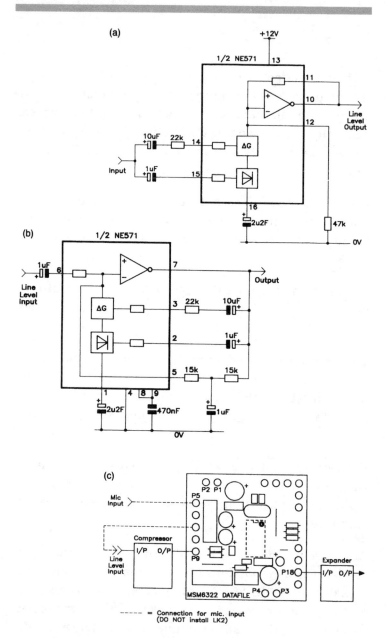

Figure 3.23   Simple compander circuit

MSM6322 parts list

Resistors — All 0.6 W 1% metal film

R1	820 Ω	1	(M820R)
R2,4	100 k	2	(M100K)
R3	1 k	1	(M1K)
R5,6	10 k	2	(M10K)
RV1	100 k vert encl preset	1	(UH19V)

Capacitors

C1,4	220 µF 10 V min electrolytic	2	(JL06G)
C2,3	100 nF 16 V minidisc	2	(YR75S)
C5	10 µF 16 V min electrolytic	1	(YY34M)
C6,8	220 nF polyester	2	(BX78K)
C7,9, 12,13	33 pF ceramic	4	(WX50E)
C10,11	100 µF 10 V minelect	2	(RK50E)

Semiconductors

IC1	MSM6322GSK	1	(UL76H)

Miscellaneous

P1–19	1 mm PCB pin	1	(FL24B)
XT1	MP crystal 4 MHz	1	(FY82D)
	printed circuit board	1	(GE78K)
	constructors' guide	1	(XH79L)

167

SSM2120 dynamic range processor

The SSM2120 is a versatile integrated circuit designed for the purpose of processing dynamic signals in various analogue systems including audio. This dynamic range processor consists of two voltage controlled amplifiers and two level detectors. These circuit blocks allow the user to logarithmically control the gain or attenuation of the signals presented to the level detectors depending on their magnitudes. This allows the compression, expansion or limiting of a.c. signals which are some of the primary applications for the SSM2120. The device will operate over a wide range of power supply voltages between ±5 V and ±18 V. Figure 3.24 shows the integrated circuit pin-out and Table 3.12 shows some typical electrical characteristics for the device. Figure 3.25 shows the integrated circuit block diagram.

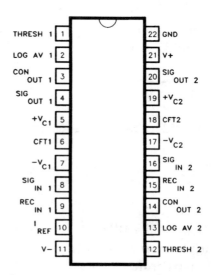

Figure 3.24   Integrated circuit pin-out

Parameter	Conditions	Min	Typ	Max
Supply voltage range				
Dual supply		±3 V	±15 V	±18 V
Single supply		+6 V	+30 V	+36 V
Supply current				
Positive			8 mA	10 mA
Negative			6 mA	8 mA
Level detectors				
Dynamic range		100 dB	110 dB	
Input current range		30 nA_{P-P}		3 mA_{P-P}
Output offset voltage			±0.5 mV	±2 mV
Frequency response				
	$I_{IN} = 1\ mA_{P-P}$			1 MHz
	$I_{IN} = 10\ \mu A_{P-P}$			50 kHz
	$I_{IN} = 1\ \mu A_{P-P}$			7.5 kHz
VCAs				
Frequency response	Unity-gain or less			250 kHz
Control feedthrough				
(trimmed)	$R_{IN} = R_{OUT} = 36\ k\Omega,$			
	A_V 0 dB to −30 dB		750 μV	
Gain control range	Unity gain	−100 dB		+40 dB
THD (unity gain)	+10 dBV IN/OUT		0.005%	0.02%
Noise (20 kHz bandwidth)	Ref: 0 dBV		−80 dB	

Table 3.12 Typical electrical characteristics

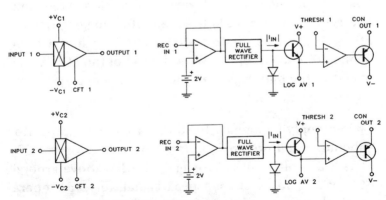

Figure 3.25 SSM2120 block diagram

Circuit description

The SSM2120 effectively contains two duplicate parts; each with a level detector and a voltage controlled amplifier.

Level detector circuit

Two independent level detection circuits are provided, each containing a wide dynamic range full-wave rectifier, logging circuit and a unipolar drive amplifier. These circuits will accurately detect the input signal level over a 100 dB range from 30 nA to 3 mA peak-to-peak.

Level detector theory of operation

Referring to the level detector block diagram of Figure 3.26, the REC_{IN} input is an a.c. virtual ground. The next block implements the full-wave rectification of the input current. This current is then fed into a logging transistor (TR1) whose pair transistor (TR2) has a fixed collector current of I_{REF}. With the use of the LOG AV capacitor, the output is then the log of the average of the absolute value of I_{IN}.

When applying signals to REC_{IN}, a blocking capacitor should be followed by an input series resistor as REC_{IN} has a d.c. offset of approximately 2.1 V above ground. Choose R_{IN} for a ±1.5 mA peak signal; for ±15 V operation, this corresponds to a value of 10 kΩ.

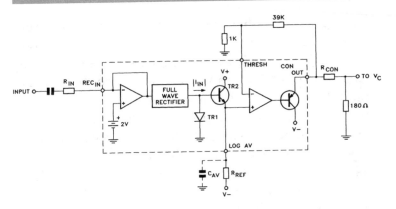

Figure 3.26 Level detector block diagram

A 1.5 MΩ value of R_{REF} from LOG AV to −15 V will establish a 10 μA reference current in the logging transistor, biasing the transistor in the middle of the detector's dynamic current range in dB to optimise dynamic range and accuracy. The LOG AV outputs are buffered and amplified by unipolar drive op-amps.

The attenuator from CON_{OUT} to the appropriate voltage controlled amplifier control port establishes the control sensitivity. Use a 180 Ω attenuator resistor to ground and choose R_{CON} for the desired sensitivity.

Voltage controlled amplifiers

The two voltage-controlled amplifiers are full Class A current in/out devices with complementary dB/V gain control ports. The control sensitivities are +6 mV/dB and −6 mV/dB. A resistor divider is used to adapt the sensitivity of an external control voltage to the range of the control port.

171

Audio IC projects

The signal inputs behave as virtual grounds. The input current compliance range is determined by the current into the reference current pin. This current is set by connecting a resistor to V+. The current consumption of the voltage controlled amplifiers is directly proportional to I_{REF}, which is nominally 200 μA, giving input and output clip points of ±400 μA. The device will operate at lower current levels, but with a reduced effective dynamic range.

The voltage controlled amplifier outputs are designed to interface directly with the virtual ground inputs of external operational amplifiers configured as current-to-voltage converters. The power supplies and selected compliance range determines the values of input and output resistors required. Note that the signal path through the voltage controlled amplifier, including the output current-to-voltage converter, is non-inverting.

Trimming the voltage controlled amplifiers

The control feedthrough (CFT) pins are optional control feedthrough null points. CFT nulling is required in applications such as noise gating and downward expansion. Applications such as compressors/limiters typically do not require CFT trimming because the voltage controlled amplifier operates at unity-gain, unless the signal is large enough to initiate gain reduction, in which case the signal masks control feedthrough. This trim is ineffective for voltage-controlled filter applications. If trimming is not used, leave the CFT pins open.

Kit available

A kit of parts is available to build several application circuits using the SSM2120. The kit includes a high quality fibreglass printed circuit board with a screened printed legend to aid construction, see Figure 3.27. Figure 3.28 shows the circuit diagram used to produce this printed circuit board. Note that, because the module may

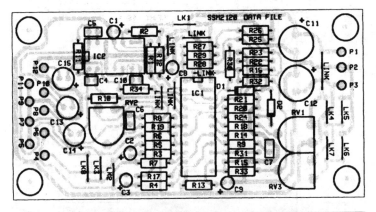

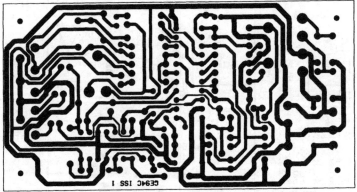

Figure 3.27   PCB legend and track

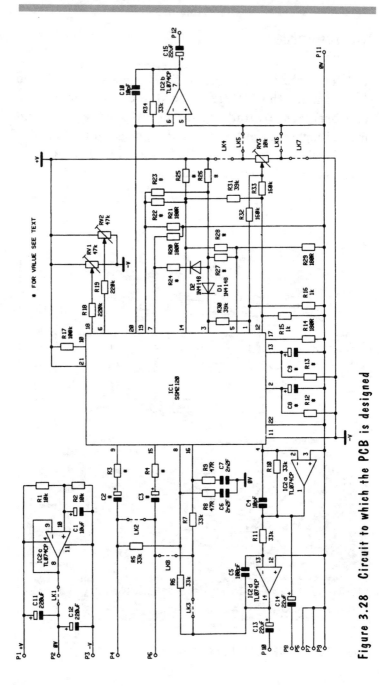

Figure 3.28 Circuit to which the PCB is designed

be used in many different applications, some of the component values supplied in the kit have been assigned an arbitrary value. For this reason minor modifications may be necessary to adapt the circuit to individual purposes.

The SSM2120 requires a split rail supply and will operate over a wide range of voltages between ±3 V and ±18 V. However, additional components have been included in the design to allow the circuit to operate from a single rail supply of between 6 V and 36 V, by installing link LK1. It is important that the supply is adequately decoupled in order to prevent the introduction of mains derived noise onto the supply rails. For optimum performance a regulated power supply should be used. All application circuits here are optimised for use with a ±15 V power supply (+30 V power supply with LK1 fitted).

The current into the reference pin determines the input and output compliance range of the voltage controlled amplifiers. This current has a nominal value of 200 µA, and is set by resistor R7; for ±15 V operation this corresponds to a value of 100 kΩ.

Figure 3.29 shows the basic wiring information.

Applications

Figure 3.30 shows the control circuit for a typical downward expander, providing a negative unipolar control output. This is typically used in downward expander, noise gate and dynamic filter applications. Here, the

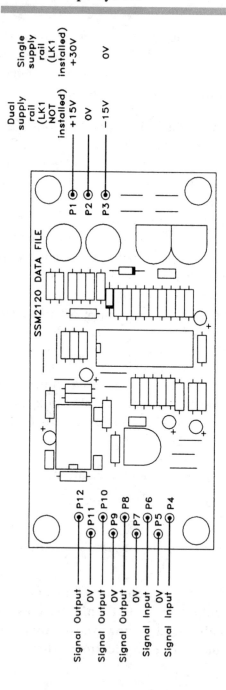

Figure 3.29 Basic wiring diagram

threshold control preset RV3 sets the signal level versus control voltage characteristics. The sensitivity of the control action depends on the value of resistor R32.

For a positive unipolar control output add two diodes, see Figure 3.31. This is useful in compressor/limiter applications.

Bipolar outputs can be achieved by connecting resistor R26 from the opamp output to V+. This is useful in compander circuits as shown in Figure 3.32. The value of resistor R26 will determine the maximum output from the control amplifier.

An attenuator resistor (R24/R27) from CON_{OUT} to the appropriate voltage controlled amplifier control port establishes the control sensitivity.

As mentioned previously, in applications such as noise gating and downward expansion the voltage controlled amplifiers require trimming as follows.

Apply a 100 Hz sinewave to the control point attenuator (D1 side of resistor R27 for voltage controlled amplifier 1, R31 side of resistor R22 for voltage controlled amplifier 2). The signal peaks should correspond to control voltages which induce the VCA's maximum intended gain and at least 30 dB of attenuation. Adjust presets RV1/RV2 for minimum feedthrough.

In all other applications, leave the CFT pins open by not fitting resistors R18 and R19.

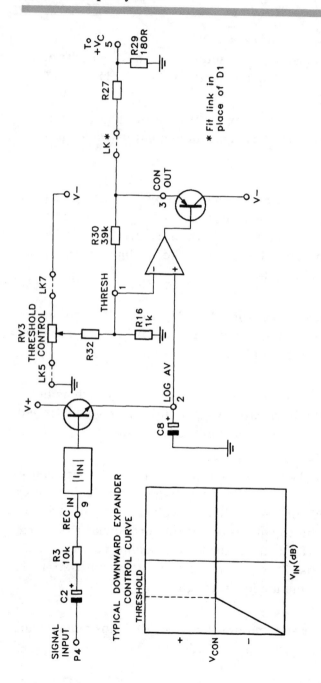

Figure 3.30 Downward expander control circuit

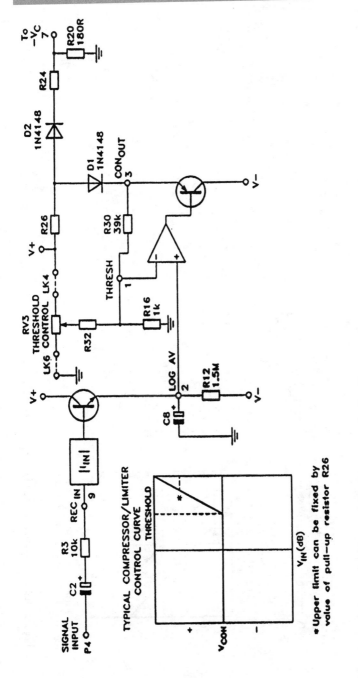

Figure 3.31 Compressor/limiter control circuit

179

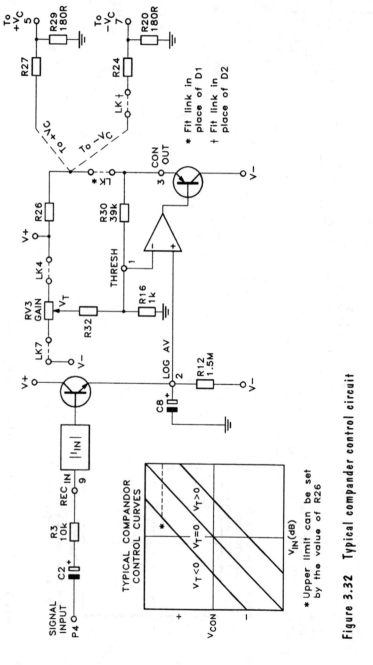

Figure 3.32 Typical compander control circuit

Companding noise reduction system

A complete companding noise reduction system is shown in Figure 3.33. Normally, to obtain an overall gain of unity, the value of resistor R24 (compression) is equal to resistor R22 (expansion). As shown in Table 3.13, the relative values of resistors R22/24 will determine the compression/expansion ratio. Note that signal compression increases gain for low level signals and reduces gain for high levels while expansion does the reverse. The exact compression/expansion ratio needed depends on the recording medium being used. For example, a household cassette deck would require a higher compression/expansion ratio than a professional tape recorder.

Compression/ expansion/ ratio	R22/24 (kΩ)	Gain (reduction or increase) (dB)	Compressor only output signal increase (dB)	Expander only output signal increase (dB)
1.5:1	11.800	6.67	13.33	22.67
2:1	7.800	10.00	10.00	30.00
3:1	5.800	13.33	6.67	33.33
4:1	5.133	15.00	5.00	35.00
5:1	4.800	16.00	4.00	36.00
7.5:1	4.415	17.33	2.67	37.33
10:1	4.244	18.00	2.00	38.00
AGC*/limiter	3.800	20.00	0	40.00

Note: *AGC for compression only

Table 3.13 Compression/expansion ratios

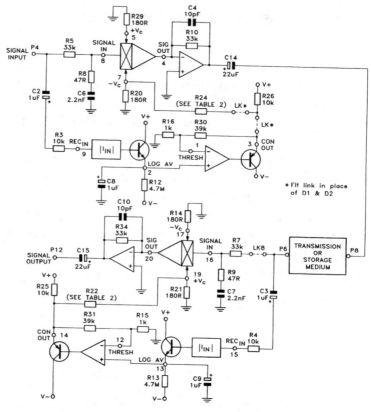

Figure 3.33 Companding noise reduction system

Dynamic filter

Figure 3.34 shows a dynamic filter capable of single ended (non-encode/decode) noise reduction. Dynamic filtering limits the signal bandwidth to less than 1 kHz unless enough *highs* are detected in the signal to cover the noise floor, when the filter opens to pass more of the audio band. Such circuits usually suffer from a loss of high-frequency content at low signal levels because their control circuits detect the absolute amount of highs present in the signal. This circuit, however, measures wideband level as well as highfrequency band level to produce a composite control signal combined in a 1:2 ratio. The upper detector senses wideband signals with a cut-off of 20 Hz, while the lower detector has a 5 kHz cut-off to sense only high-frequency band signals. Unfortunately, even in this system, a certain amount of mid- and high-frequency components will be lost, especially during transients at very low signal levels. The threshold control, preset RV3, sets the filter characteristics for 50 dB (V+) to 90 dB (V−) dynamic range programme source material.

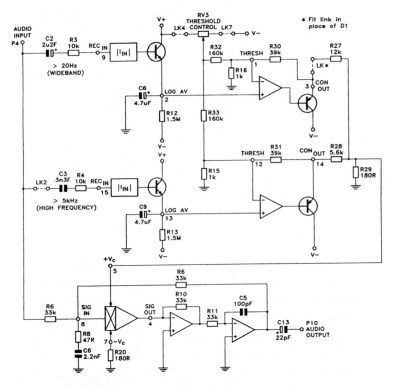

Figure 3.34 Dynamic noise filter circuit

Dynamic filter with downward expander

As shown in Figure 3.35, the output from the wideband detector can also be connected to the $+V_C$ control port of the second voltage controlled amplifier which is connected in series with the sliding filter. This will act as a downward expander with a threshold that tracks that of the filter. Downward expansion uses a voltage controlled amplifier controlled by the level detector. This section maintains dynamic range integrity for all levels

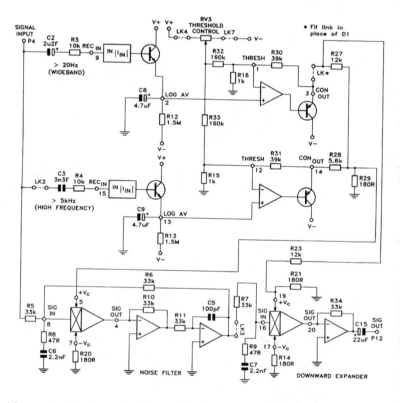

Figure 3.35   Dynamic filter with downward expander

above the threshold level (set by preset RV3) but, as the
input level decreases below the threshold, gain reduction occurs at an increasing rate, as shown in Figure 3.36.
This technique reduces audible noise in fadeouts or low
level signal passages by keeping the standing noise floor
well below the programme material. Using this system,
up to 30 dB of noise reduction can be realised while pre-
serving the crisp highs with a minimum of transient side
effects.

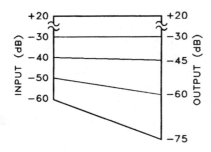

Figure 3.36   Typical downward expander I/O characteristics at
−30 dB threshhold level (1:1.5 ratio)

Companding noise reduction system parts list

Resistors — All 0.6 W 1% metal film (unless specified)

R1,2,3,4, 25,26	10 k	6
R5,7,10,34	33 k	4
R8,9	47 Ω	2
R12,13	4M7	2
R14,20, 21,29	180 Ω	4
R15,16	1 k	2
R17	100 k	1
R18,19	220 k	2
R22,24	see table 2*	2
R30,31	39 k	2
RV1,2	47 k hor encl preset	2

Capacitors

C1	10 µF 16 V min elect	1
C2,3,8,9	1 µF 63 V min elect	4
C4,10	10 pF ceramic	2
C6,7	2n2F ceramic	2
C11,12	220 µF 35 V PC electrolytic	2
C14,15	22 µF 35 V min elect	2

Semiconductors

D1,2	fit link	2
IC1	SSM2120P	1
IC2	TL074CN	1

Miscellaneous

P1–12	1 mm PCB pins	12
LK8	fit link	1

*Note resistors R22 and R24 are supplied as 7k5 in the kit.

Dynamic noise filter with downward expander parts list

Resistors — All 0.6 W 1% metal film (unless specified)

R1,2,3,4	10 k	4
R5,6,7,10, 11,34	33 k	6
R8,9	47 Ω	2
R12,13	1M5	2
R14,20, 21,29	180 Ω	4
R15,16	1 k	2
R17	100 k	1
R18,19	220 k	2
R23,27	12 k	2
R28	5k6	1
R30,31	39 k	2
R32,33	160 k	2
RV1,2	47 k hor encl preset	2
RV3	10 k hor encl preset	1

Audio IC projects

Capacitors

C1	10 µF 16 V min elect	1
C2	2 µ2F 63 V min elect	1
C3	3n3F ceramic	1
C5	100 pF ceramic	1
C6,7	2n2F ceramic	2
C8,9	4 µ7F 35 V min elect	2
C11,12	220 µF 35 V PC electrolytic	2
C15	22 µF 35 V min elect	1

Semiconductors

D1	fit link	1
IC1	SSM2120P	1
IC2	TL074CN	1

Miscellaneous

P1–12	1 mm PCB pins	12
LK2,3,4,7	fit link	4

Dynamic noise filter parts list

Resistors — All 0.6 W 1% metal film (unless specified)

R1,2,3,4	10 k	4
R5,6,7,10,11	33 k	4
R8	47 Ω	1
R12,13	1M5	2

R20,29	180 Ω	2
R15,16	1 k	2
R17	100 k	1
R19	220 k	1
R27	12 k	1
R28	5k6	1
R30,31	39 k	2
R32,33	160 k	2
RV2	47 k hor encl preset	1
RV3	10 k hor pncl preset	1

Capacitors

C1	10 µF 16 V min elect	1
C2	2µ2F 63 V min elect	1
C3	3n3F ceramic	1
C5	100 pF ceramic	1
C6	2n2F ceramic	1
C8,9	4 µ7F 35 V min elect	2
C11,12	220 µF 35 V PC electrolytic	2
C13	22 µF 35 V min elect	1

Semiconductors

D1	fit link	1
IC1	SSM2120P	1
IC2	TL074CN	1

Miscellaneous

P1–12	1 mm PCB pins	12
LK2,4,7	fit link	3
Magazine Version		

SSM2120 parts list

Resistors — All 0.6 W 1% metal film (unless specified)

R1,2,3,4, 25,26	10 k	6	(M10K)
R5,6,7, 10,11.34	33 k	6	(M33K)
R8,9	47 Ω	2	(M47R)
R12,13	4M7	2	(M4M7)
R12,13	1M5	2	(M1M5)
R14,20,21,29	180 Ω	4	(M180R)
R15,16	1 k	2	(M1K)
R17	100 k	1	(M100K)
R18,19	220 k	2	(M220K)
R22,24	see table 2 (nominally 7k5)	2	(M7K5)
R23,27	12 k	2	(M12K)
R28	5k6	1	(M5K6)
R30,31	39 k	2	(M39K)
R32,33	160 k	2	(M160K)
RV1,2	47 k hor encl preset	2	(UH05F)
RV3	10 k hor encl preset	1	(UH03D)

Capacitors

C1	10 µF 16 V min elect	1	(YY34M)
C2,3,8,9	1 µF 63 V min elect	4	(YY31J)
C2	2µ2F 63 V min elect	1	(YY32K)
C3	3n3F ceramic	1	(WX74R)
C4,10	10 pF ceramic	2	(WX44X)
C5	100 pF ceramic	1	(WX56L)

C6,7	2n2F ceramic	2 (WX72P)
C8,9	4µ7F 35 V min elect	2 (YY33L)
C11,12	220 µF 35 V PC electrolytic	2 (JL22Y)
C13,14,15	22 µF 35 V min elect	3 (RA54D)

Semiconductors

D1,2	1N4148	2 (QL80B)
IC1	SSM2120P	1 (UL78K)
IC2	TL074CN	1 (RA69A)

Miscellaneous

P1–12	1 mm PCB pins	1 (FL24B)
	printed circuit board	1 (GE94C)
	constructors' guide	1 (XH79L)

MAPLIN Books

This book is part of a new series developed by Butterworth-Heinemann and Maplin Electronics. These practical guides will offer electronics constructors and students a clear introduction to key topics. The books will also provide projects and design ideas; and plenty of practical information and reference data.

0 7506 2053 6 **STARTING ELECTRONICS**

0 7506 2123 0 **COMPUTER INTERFACING**

0 7506 2121 4 **AUDIO IC PROJECTS**

0 7506 2119 2 **MUSIC PROJECTS**

0 7506 2122 2 **LOGIC DESIGN**

These books are available from all good bookshops, Maplin stores, and direct from Maplin Electronics. In case of difficulty, call Reed Book Services on (0933) 410511.